T0123537

QUARKS
TO
COSMOS

Science as Hierarchy. Humans in the Hierarchy.

J. MAILEN KOOTSEY PH.D.

authorHOUSE®

AuthorHouse™
1663 Liberty Drive
Bloomington, IN 47403
www.authorhouse.com
Phone: 833-262-8899

Published by AuthorHouse 07/09/2021

ISBN: 978-1-6655-2970-9 (sc)
ISBN: 978-1-6655-2968-6 (hc)
ISBN: 978-1-6655-2969-3 (e)

Library of Congress Control Number: 2021912603

Print information available on the last page.

Dedication

For Lynne, Brenden, and Sean.

CONTENTS

PREFACE

The Sabbath Seminars book discussion group at Loma Linda University substantially contributed to this book by dedicating five of their two-hour sessions to reviewing early drafts. The vigorous discussions of this group were invaluable to me by helping me clarify explanations of ideas in several areas. In particular, I would like to especially thank Jim Walters, Sabbath Seminars leader, for scheduling the sessions for my book and for his helpful suggestions, Jan Long for his detailed review as a scientific lay person, Lucerne French for help with grammar and word selection, Ralph McDowell for keeping my rambling thoughts within the bounds of philosophical logic, and Dave Hessinger for his scientific critique.

Individuals outside the Loma Linda University community also were of significant assistance. Ronald L. Numbers and his publications helped me relate my book to the long line of existing science and religion books. Richard Coffen, a long-time friend and retired book editor, gave me many helpful suggestions ranging from sentence structures to clarity of explanations. From Andrews University, Shandelle Henson provided valuable tips for my mathematics examples, and Jim Hayward pointed me toward additional biological references.

FOREWORD

If we were to toss a fair coin, say 100 thousand times, and record the outcome of each event—heads or tails—by the time we got to the 100 thousandth toss we would have recorded 50% head, and 50% tails. We know this from the laws of probability. While this overall outcome has a determinate characteristic about it, amazingly each individual trial is indeterminate. For example, if, in the process of the coin toss we ran into a pattern of 10 straight heads in a row, some might be inclined to speculate that the 11th toss will have an increased probability of producing tails. However, those who might make this assumption would be mistaken for each trial is independent of the next, which means that heads on the 11th toss is just as likely as tails.

Physicists have discovered that physical reality works sort of like this as well. The macro scale evidences a predictable, determinate pattern, while the quantum scale is indeterminate. This has led the search for a unifying principle that could bridge this divide.

Until this riddle is resolved, our author, Mailen Kootsey, seeks an interim solution by putting forward several important unifying concepts. We can even speculate that once a fully unifying principle is found, that the ideas laid out in this book may be seen as even more integral to our understanding than what is currently appreciated.

I will mention three of the concepts he puts forward. First, he proposes a comprehensive organization that he refers to as the *Catalog of the Universe.* From this "Catalog" framework, our author seeks to capture everything—including the identification of real physical entities in the universe and other possible entities that have existed or could exist. We are talking here about objects from the very large to the very small; from the physical to the social sciences, and to the ethereal realm of religion as well.

The second concept involves recognizing these entities (also known as holons), as objects that form hierarchies. These hierarchies are identifiable as levels that are connected to each other, that run from simple to complex, from the very small to the very large, and from the random to certain specified orderings, such as that which we find in biology. The dimensionality of the hierarchy can be determined either by its size or the complexity of objects found at different levels.

A third concept he introduces is that this complexity hierarchy is connected by relationships, that then form networks to fill the universe. We could think of this all being a worldview approach because of its strategic big-picture perspective.

To set the stage for some of these ideas in a more concrete way, it may be helpful to start our thinking in the basement of physical reality, where we find a quantum world of subatomic particles that are intimately connected with each other in ways that form the atom at a higher level. Initially it was believed that the atom was the basement of matter, yet now we know that even the atom has subparts that relate to each other. In fact, the atom is made possible by the relationships formed by these sub-particles. Atoms, in turn, are the foundational building blocks for molecules, cells, and larger structures we are more familiar with at a higher sensate level.

At the upper end of the physical spectrum, where matter exists on a cosmic scale, we observe once again a physical hierarchy as well as relationship patterns, with comets, asteroids, and planets relating to a common star in a solar system, where moons relate to planets and solar systems relate to galaxies.

In between the two ends of the size-scale are other levels, and from them all—from quantum, middle, to cosmic scale—come all the objects in the material universe, that relate to each other in a grand networked architecture. One of these *middle* structures that is found in this hierarchy of objects is biology which has its own subset of hierarchical levels, combined with relationships and networks. At the minute end, we find four unique nitrogenous bases that act in relationship to each other, forming a digital code that makes life possible. It is the unique combinations of all these coded relationships along with emergent complexity that produces bio-systems, as well as the biodiversity that we now observe. The end products of this coding are organisms, populations, and ecosystems.

At the human level we do have some experience with aspects of this, for as a species we operate in relationship to, and are often networked with each other. Most of the time we don't even think about it — it just happens in the context of families, schools, churches, special interest associations, and the power structures of industry and government. For many, the concept of *relationship* is primarily about distinctly human social and legal interactions. But as we have discussed above, our author points out that the concept of *relationship* is pervasive throughout the physical universe. While it has been said, "No man is an island," the author's point is that there are no true islands given the pervasive role of relationships and networks driving the order of all things in the universe.

As a physicist with a career in academia, Kootsey, our author, is attempting in this book to bring to the reader contemporary ideas from science, and he warms up the discussion by weaving in personal anecdotes. Readers will find the book readable and informative, showing how the entire material universe is built on hierarchies, relationships and networks. Ultimately the ebb and flow of these structures, and their complexities form one unified structure of reality that makes up our cosmos.

Thinking in these terms has strategic consequences for scientific inquiry, as it requires thinking in more holistic terms. It also has practical implications for everyday life beyond science as we consider organizing the world around us. The worldview it provides suggests that everything we do for good or ill ripples out in some way to the larger world. We are, after all, a part of the furniture of the universe (to use a Stuart Kauffman framing), with the capacity to proceed in positive or negative ways.

For many years I have been fascinated by the power of science to explain the natural world by combining observation of regularities with reason. While all these concepts—*hierarchies, relationships* and *networks*—are floating around the sciences in various ways, no one has actually attempted to bring these concepts together as a way of unifying physical and nonphysical reality. Even though these concepts have never been joined this way, we will likely recognize them to be concepts that play an integral role by operating as a subtext of most every idea or discussion. So, our author has collected these concepts together, and is attempting to fill in a part of the unification puzzle.

In the end, we may ask ourselves how this grand architecture came about?

For a long time, people of faith have had a story tradition that explained it all, and while systematic observations of physical reality have affirmed some of these traditions, other observations have not always been affirming. This has caused friction between the community of faith and that of science, leading some to distrust science. In a 2003 book, biologist Stephen Jay Gould made famous the NOMA concept, an acronym for "non-overlapping magisteria", suggesting that science legitimately operates in one realm, while theology operates in its own realm. It was an attempt to find peace between science and religion, and many have found it to be a useful distinction. Kootsey acknowledges this approach to be helpful, but he remains consistent to his main thesis by reminding readers that science and religion should not be seen as categorically distinct magisteria, but rather differing levels of the complexity hierarchy that necessarily involve differing methodologies—ultimately both being part of one unified structure.

I have been acquainted with the author, Mailen Kootsey, for a number of years and have come to deeply respect his understanding of reality as an academic and as a physicist. I have found him to be grounded and thoughtful in his articulations. His strategic view of the world is on full display in this book and is well worth reading for those who want to have a little better understanding of how different aspects of reality fit together. To this end, readers should find this book is an important part of the conversation.

Jan M. Long
Director of Institutional Research
La Sierra University

PROLOGUE

The Making of a Multidisciplinary Scientist

> All are but parts of one stupendous whole, Whose body Nature is, and God the soul. *Alexander Pope.*

I have always been curious about how things work. As far back as I can remember, I delighted in taking things apart to study their inner workings. I learned about gears and levers, about electricity and bulbs, rocks, bugs, stars, and electronics. The latter especially fascinated me, and as an early teenager, I learned to build simple radios and repair commercial radios, televisions, and other home appliances. I learned Morse code and passed the tests to become a radio amateur or "ham."

In the fall of 1956, I planned to enroll at Pacific Union College (PUC), a small liberal arts college in the wine country of northern California. It was necessary to pick a major subject, and one of the natural sciences would have been an obvious choice. In the 1950s, liberal arts colleges like PUC offered just three natural science majors: biology, chemistry, and physics. Chemistry appeared to be a discipline of naming molecules and learning to follow recipes, although there was a course called Physical Chemistry that students seemed to dread. Biology also emphasized naming and classification. Neither seemed to fit my curiosity, so I chose physics -- the subject focused on mechanisms.

All three of the sciences offered presumably believed in the same scientific method. The human study of the natural world since the 16th century has consisted of a combination of experiments with theories based on the collected data. So why was there a significant difference between teaching content in the three sciences? It would take more education and years working in science to answer that question.

In physics, theories meant mathematics. Entering college, I was blissfully ignorant about the kinds of mathematics needed for scientific models. The only real mathematics taught in high schools at that time was algebra. Since my high school algebra class was weak, I took a correspondence course in college algebra during the summer before enrolling in college. I did not know then that 2500 years earlier, the Greeks were frustrated because the forms of mathematics they knew – algebra and geometry – were inadequate to describe changing processes they observed in the real world. Two thousand years later, new mathematics of change filled the need: calculus. If I wanted to be a scientist, I would need to learn calculus.

At the end of my freshman year at PUC, the Physics Department Chairman asked me if I would consider staying on campus during the summer to work in the Department repairing and updating physics laboratory equipment. The small college town in northern California was a pleasant place to be in the summer. The work would be to my liking, so I agreed to take on the job – with one condition. I knew I must take calculus in my sophomore year 1, so I asked if the senior mathematics professor would teach me calculus during the summer, knowing that he was one of the finest teachers on campus. Happily, he agreed, and we spent hours together several days a week for the summer term while I worked on physics equipment. Together we went through the entire calculus textbook, with me working every exercise in the book on the blackboard, aided occasionally by this master teacher. By the end of the summer, my dread of this mathematical subject had changed to appreciation. Calculus was the tool that later made it possible for me and other scientists to begin to understand how chemical reactions and biological systems work.

In my sophomore year at college, another unique opportunity came my way. The Physics Department at PUC purchased an early digital computer. In 2021 nobody would notice this event, but in 1957, there were no smartphones, tablets, or home computers. Even many large corporations did not have a computer yet, so to have one at a small private college was a rare opportunity. I knew that some scientific and engineering calculations required thousands of simple mathematical calculations because of a previous summer job. This job was with a large engineering company that was designing nuclear reactors for electrical power generating plants.

Along with several other students, my task was to perform a long series of simple arithmetic calculations using an electrically powered mechanical desk calculator. Each step was simple, but a series lasted for thousands or tens of thousands of them. The accuracy had to be perfect at each step because one error would ruin the entire series – sometimes a whole day's work! Based on that experience, I was ready for a machine that could do arithmetic rapidly and without errors.

The Bendix G15-D computer purchased by PUC was about the size of two refrigerators, weighed 966 pounds, and its electronics required 450 vacuum tubes. [2] The computer broke down about once a day. Since no service contract was available, the Department faculty, staff, and students had to learn how to repair it. It had only one kind of memory – a magnetic drum holding a little over 2000 20-bit words, or in today's language 0.01 megabytes, and it could do 400 multiplications per second. Programming this computer had to be done in binary as there was no high-level programming language. The G-15D was a very primitive and slow computer, but it introduced me to digital computing. I now had a way to do thousands of tedious calculation steps quickly and without errors. Computing and calculus turned out to play significant roles in my career. These two tools changed the way to study scientific theories.

After graduating from college, I drove from California to enter the graduate program in physics at Brown University in Rhode Island. It was a requirement for graduate students at Brown to earn a master's degree before starting a doctoral research project. I learned that Brown had a well-equipped computer facility because the University had done a personal favor for Thomas J. Watson Sr., an early chairman and CEO of IBM. I petitioned the Physics faculty for permission to do a computer simulation for my master's project – replacing the typical experiment. Substituting a computer simulation for a physical investigation had never been done before at Brown and possibly any university, but the Physics faculty agreed. I completed the project and received my master's degree in 1964. I had to do some research for my doctoral degree: building most of the necessary equipment and rejuvenating the available accelerator. My earlier summer job working on physics laboratory equipment was just the preparation needed for that project.

Toward the end of my research and thesis writing at Brown, my thoughts turned to career alternatives. I could join an experimental research group at a national laboratory, following my interest in designing and building research equipment. I could try for a job at a research university where my responsibilities would combine research and teaching. Or I could choose to teach at a small college like the one I attended, where I might be able to do a little research on the side. I could even leave the academic world and find a job in industry.

Before I narrowed my search down to one of these traditional career alternatives for a physics graduate, a different opportunity unexpectedly became available. My former Physics Department Chair at PUC called, telling me that he had moved to the Physiology and Pharmacology Department at Loma Linda University Medical School in Southern California. Would I be interested in joining him there as a postdoctoral student to learn how to apply physics thinking to medical challenges? The Bank of America Giannini Foundation would grant a fellowship to pay my salary for two years. My wife was pleased with the thought of returning to California, where she grew up, so back to the West we went from Rhode Island.

My task for the next four years – two paid by the Giannini fellowship and two more paid for by Loma Linda University Medical School – was to learn medical physiology by apprenticeship. I sat in on classes with medical and graduate students and assisted with some of the Department's research projects. In the laboratories for medical students, I learned dog surgery. And I read a lot: medical physiology books and research papers in several specialty areas of physiology. I awoke to the idea that it was possible to use physics and chemistry principles to think about how living systems work. By the end of the second year, it was clear in my mind that I wanted to pursue a career in some branch of physiology where my physics, computing, and instrument design experience would come together.

I selected cardiac electrophysiology as a research area: electrical activity of heart muscle triggering the contraction that pumps blood. No one at Loma Linda was studying that area, so I needed to find a productive research group at another university. More hours reading physiology research papers narrowed my focus down to three campuses. I also discovered that the National Institutes of Health (NIH) concluded that more individuals from the physical sciences should be working on biomedical mechanisms. They offered two years of funding to individuals from the physical sciences to transition into a medically related field: a program designed just for me! I applied and received one of the fellowships – providing that I could find a suitable research laboratory that would accept me.

After spending many hours reading research papers in cardiac electrophysiology, one author's name stood out: Dr. Madison S. Spach at Duke University Medical Center, a pediatric cardiology group leader. I contacted Dr. Spach, explaining my background and interest and the two-year fellowship. Would it be possible for me to visit him at Duke? Yes, he said, and I promptly arranged to fly to North Carolina. On arriving at the Duke campus, I searched out Dr. Spach's office – no easy task in the rambling 1000 bed Duke Hospital. His office was tiny; it seemed like the proverbial converted broom closet with every shelf and horizontal space piled with papers and books. But Dr. Spach greeted me with southern hospitality, and he cleared off a chair for me to sit. He had a simple message for me: we would like to have you at Duke, but rather than working in my laboratory, said Dr. Spach, I should go to the Physiology Department and work with Dr. Edward "Ted" Johnson's group in the Cardiac Electrophysiology Lab. I took Dr. Spach's advice and worked with both Drs. Spach and Johnson. They were co-leaders of a multidisciplinary team, something new for me and, as I learned later, something new at that time for scientific research in general. I eventually spent 20 years working with the Duke team on heart electrophysiology and the electrocardiogram.

In 1989 Gustavus Adolphus College in Minnesota held a symposium titled "The End of Science?" The organizers' intended premise was that belief in science, not science itself, was coming to an end. One of the speakers was Gunter Stent, a biologist from the University of California at Berkeley. Twenty years before the symposium, Stent had written a book entitled *The Coming of the Golden Age: A View of the End of Progress,* predicting not that science would run out of new insights but rather that progress would accelerate exponentially until science collapsed. Seven years later, popular science writer John Horgan published his book *The End of Science: Facing the Limits of Knowledge in the Twilight of the Scientific Age.* He expanded on Stent's prediction by predicting that progress was nearing an end in eight major science sub-fields.

Science progress did not end in the 20th century, of course. New laboratory techniques reinvigorated experimental science, and in this book,

I describe two advances in theoretical methods that gave new life to theories.

By the accident of when I was born, my education and research career occurred precisely when scientific research was making those significant changes in methodology. Science was transitioning from isolated fields to a hierarchy of interlinked concentrations through multidisciplinary teams. In this book, I call this hierarchy the Catalog of the Universe. Chapters 2 through 5 define the Catalog structure and its properties, explain its name, and illustrate how it provides a unified view of science. A single principle underlies all levels in this hierarchy. As I will explain, the Catalog includes humans as biological creatures and in cooperative groups doing things like education, philosophy, science, and religion. I compare the goals and research methods used at different levels of the hierarchy in Chapters 6 and 7.

Mathematics provided essential tools for building scientific theories starting before the ancient Greeks. At the beginning of the 16th century, the available mathematics consisted of algebra and geometry, developed by ancient Arabs and Greeks. These systems were suitable for static phenomena but inadequate once it was recognized how universal change is. A new mathematics system for studying time-changing systems was developed late in the 17th century and is described in chapters 8 through 10.

Another change in science research began in the middle of the 20th century and resulted from the availability of fast digital computing. I have already hinted that computers could automate repeating many tiny calculation steps, but the change theoretically involves more than speed. Science usually consists of repeated interactions between experiment and theory. The availability of rapid calculations created a new branch of science: experimental theory. Chapters 11 through 14 explain this recent activity.

Chapters 15 through 18 describe the Catalog hierarchy's applications to human social interactions and religion, including showing their relationships to studies conventionally regarded as "sciences." Scientists do their work of discovery to benefit humans, but this work is not always appreciated by all, as discussed in Chapter 19. Chapter 20 celebrates the human benefits from science and recent improvements in research methods.

CHAPTER 1

Humans and Their Universe Before Science

> There is no culture recorded in human history which has not practiced some form of religion. *Joshua J. Mark.* [1]

We humans live in a complex environment: the ground below, the surrounding air, nearby plants, animals, other humans, and the sky above with lights bright and faint. Continued survival requires an understanding of this environment. Which aspects help us stay alive and comfortable, and which others would bring pain and death if encountered? We have a brain enabling learning about all these things and thinking about our relations to them. What solutions have we found over the centuries to relate to the universe?

Western science in the 21st century combines observations of the physical world with mathematical theories to understand natural phenomena for humans' benefit. Humans practiced this science for only a tiny fraction of the whole span of their existence, beginning in the 16th century CE. Like a living organism, science did not start fully mature but grew both in size and methodology. This book describes critical changes in science practice since the middle of the 20th century that have greatly increased research productivity and understanding. As background for describing recent developments, this chapter is a brief review of how humans related to their world before modern science.

From the earliest humans to the time of the Greeks – about the 5th century BCE -- people of the Western civilization believed that gods with human personalities and supernatural powers [1] controlled everything in the environment: the weather, the seasons, the fecundity of wives and herds, the growth of crops, and the results of battles. Gods were believed to argue and fight with each other, die, and come to life again to fight some more. These gods' numbers and names varied from one human group to another, but the characteristics were the same. Priests representing the gods told the humans what to do to keep their gods happy and doing good things for humans, but sometimes it seemed hard to please these gods.

Greeks believed in gods, but they also began to think about how things work in the natural world. What were objects and humans made of? Were there regularities in the movement of objects, for example, that laws could describe? As thinking tools, the Greeks had the mathematics systems of algebra and geometry, but these were inadequate to describe what they observed in Nature, especially when there were changes with time.

Plato wrote that things on earth were imperfect copies of perfect Ideal objects in a supernatural realm. This realm also contained the One, a perfect being or Ideal of Ideals, responsible for creating the Ideals. Over the next few centuries, Plato's concepts developed into a philosophy that gradually replaced the earlier belief in controlling gods. This new philosophy was called The Great Chain of Being. [2] It consisted of a hierarchy with God at the top (based on Plato's Ideal of Ideals) and everything that God created in descending order below: first angels, then kings and rulers followed by lesser rulers, ordinary people, the poor and needy, domestic animals, wild animals, birds, insects, plants, and finally rocks at the bottom. God, being perfect, was one hundred percent "soul," and each successively lower level represented less and less "soul" down to rocks with zero percent "soul."

Figure 1.1
1579 drawing of the Great Chain oif Being
from Didacus Valades, *Rhetorica Christiana*

The Great Chain of Being governed Western civilization and influenced every aspect of human life, including society, religion, literature, and the arts. A perfect God must have created a perfect universe, so God created everything possible. Nature was viewed as static and unchanging since creation, except for the regular seasons. Here are some of the concepts that resulted from the Chain of Being:

- The universe is fundamentally static.
- No new species of life will appear.
- Each human is created at a stratum of society and cannot rise above that stratum.

- Humans have a dual nature: a lower, physical body plus a Soul from God.
- Evil and predation must exist to make creation complete (a theodicy).
- Life exists on other planets or worlds.

Individuals like Augustine of Hippo, Anselm, Roger Bacon, and Thomas Aquinas studied the natural world, for example, the movement of planets and stars. They built their theories of how things worked on Plato's philosophy of Ideals and the Great Chain of Being. Individuals who constructed those theories were called "natural philosophers." Authentication by the predominant religion supported the explanations, and they held sway for many hundreds of years.

But there were problems. For example, observations were beginning to make it clear that the world is not static after all. Change is everywhere, and natural philosophy provided no basis for developing theories for changing systems.

In the 16th century, a few creative individuals started building theories about the world based on their observations rather than on philosophy handed down from Plato. The next chapter begins the new science story, emphasizing evolving methods in the last hundred years.

CHAPTER 2

The Sciences Grow and Connect

The endeavour is to link and harmonise the achievements of individual investigators in their various fields of science.
Sahotra Sarkar.

By the 16th century CE, explorers were discovering fossils of creatures that no longer existed and other signs of change in the natural world. Perhaps the world was not static after all. The influence of the Great Chain of Being waned from the 16th to the 18th centuries. A new way of understanding the world was beginning to develop. Explanations no longer arose out of philosophy but instead were based on observations of the world. A search began for fundamental principles that could account for observations and results of experiments. As the new science developed through much of the 20th century, the result was a collection of isolated fields of study, independently following the new scientific method. Sharing was rare. This chapter will show how the new sciences relate to and support each other.

In the 16th and 17th centuries CE, curious individuals began compiling detailed records of observations, and they thought about principles responsible for what they saw. Their names are now familiar in the history of science: Galileo, Newton, Kepler, Copernicus, Brahe, Bacon, Descartes, Hooke, Harvey, and Boyle – to name a few. These names are associated with specific areas of science, like Newton with gravity or Harvey with blood circulation. But these explorers were polymaths, familiar with and contributing to multiple areas of study. For example, Newton originated the idea of gravity but was also a co-inventor of calculus – a new type of mathematics that was essential for the continued development of science. Harvey mapped the circulation of blood and also made significant contributions to the knowledge of human anatomy and embryogenesis.

In the 17th to 19th centuries CE, science emphasized detailed observations in Nature, compiled into large volumes for adults and children. One example is *Das Buch der Natur* by Konrad von Megenberg, published in 1861 and announced as the first guide to Nature in the German language. This book contains 442 pages of text and illustrations describing animals, birds, and plants, with another 440 pages of footnotes and indexes. Some theory contributions also came out of this period, such as the modern system of naming biological organisms by Linnaeus and Darwin's theory of evolution.

By the first half of the 20th century CE, an explosion of specialized scientific study areas had developed based on the new approach, each one focusing on a different sphere of phenomena. Physicists studied atoms and fundamental laws of motion and forces of gravity, electricity, and magnetism. Biologists were busy classifying plants and animals, from microscopic to dinosaurs. Chemists identified molecules and observed reactions between them. Physicians and physiologists were trying to decipher the functions of body organ systems. Geologists studied the properties of rocks and formations. Astronomers expanded their catalog of heavenly bodies, classified them, and theorized about their origins.

The rapid growth of scientific information produced independent scientific communities, each studying a narrow range of natural phenomena. Polymath individuals no longer dominated. Science was not a continent but rather an archipelago. Each of these communities claimed a defined territory of observations for its study and the right to name the things they studied. Each field developed the new instruments needed to enhance their studies. Relationships between these research communities were minimal or nonexistent. For example, theories in each area were independent of neighboring sciences. Education was established separately for each field with separate university departments, courses, and degrees.

The change in the basis for theories about the world from philosophy to observations was so crucial that the old name "natural philosopher" for researchers needed replacement. On June 24, 1833, the British Association for the Advancement of Science held its third meeting at the University of Cambridge. The poet and philosopher Samuel Taylor Coleridge stood up the first night of the conference and announced, "You must stop calling yourselves natural philosophers." He explained that true philosophers like himself pondered the cosmos from their armchairs. They were not mucking around in the fossil pits or conducting messy experiments with electrical piles like British Association members. Young William Whewell suggested that by analogy with the name "artist," the term "scientist" would be more appropriate for Association members. Thus the very word "scientist" emphasizes the critical role of observations in current studies of the natural universe, replacing the earlier philosophical arguments.

Did the rise of the new science make philosophy obsolete? No, says contemporary philosopher and writer Rebecca Goldstein. The sciences were dependent on philosophy

> ". . .until the scientific enterprise matured sufficiently to be able to transform them [physics, cosmology, and biology] into sciences. Then it was the turn of psychology, and then linguistics, to remove themselves from philosophy's domain and reinvent themselves as sciences. And so it has gone: the scientific enterprise transforming philosophy's airy-fairy speculations into a form that allows reality to tell us when we're wrong—right down to our own scientifically explosive period, when the advancement of cognitive and affective neuroscience and evolutionary psychology have moved human Nature itself firmly into the orbit of science. . . . Philosophy lives only to be made obsolete by science. Philosophy's role in the business of knowledge is to send up a signal reading *Science desperately needed here*. . .the real point of philosophy is to maximize our coherence by discovering and resolving the inconsistencies we accrue as we go about trying to get our bearings in the world, which is our distinctively human project." [1]

CAN THE SCIENCES BE UNIFIED?

Because physicists studied fundamental laws of atoms, forces, motion, and energy, and because physicists expressed their theories precisely in mathematics, they privately claimed superiority over all sciences. Physicists argued that chemistry, biology, geology, and physiology must be based on atoms and fundamental physics laws. All these fields were just undeveloped branches of physics since they had no mathematical theories. As a college physics major in the late 1950s, I seldom heard this territory claim mentioned explicitly in physics lectures, but it was very present in discussions and curriculum advising.

In the early 19th century, Pierre-Simon Laplace concluded that it was theoretically possible for a "demon" to predict the future. First, note the location and momentum of every particle of matter in the universe at some time instant. Then use the known laws of motion to predict their future positions and momentums. [2] Laplace's conclusion is now known to be incorrect for several reasons from fundamental principles. But at the time he proposed it, his idea reinforced the belief that physics would explain everything.

Life scientists countered physics claims by postulating a *vital force* present in all living things. This *vital force* was assumed to be a fluid-like essence like the "soul" of the Chain of Being and not subject to physics laws. Inorganic materials supposedly had none of this *vital force* and were fundamentally different from organic and living substances. This concept was introduced in the 16th century and lasted into the 20th century. Scientists eventually abandoned the *vital force* theory because it made no predictions. As we will see later, there was an element of truth to the concept. Living and non-living substances do differ fundamentally. However, describing that difference as a "fluid not subject to physics" is not sound science.

A desire for unity of the sciences has a long history; the goal remains incomplete as I write this book. Three philosophers of Greek Ionia, a region located on the West coast of present-day Turkey, were the first to propose unity bases. Thales (624-546 BCE) believed water is the fundamental substance. For Anaximander (610-546 BCE), the essence is an indefinite or eternal limitlessness. Air was the essential substance for Anaximenes (585-525 BCE). These early ideas were a long way from atoms and molecules. The Greeks were the first to propose the monist idea that a small number of material principles recognized by observing Nature could account for everything. Physicist and historian of science Gerald Holton called this the "Ionian Enchantment."

For today's physicists, unity means linking together the fundamental governing force laws into a single theory. When I was in a physics graduate program in the early 1960s, four fundamental force laws existed: electromagnetic, strong and weak nuclear, and gravitational. Since those years, theorists combined the electromagnetic and weak nuclear forces into a single electroweak force. The link became apparent as physicists developed a theory for the Big Bang and the subsequent formation of the universe as we know it, earning Nobel prizes in 1979 for the physicists Glashow, Salam, and Weinberg who formulated the combined theory. During a very early period after the Big Bang, when the budding universe's temperature was very high (approximately 10^{15} K), the electromagnetic and weak nuclear forces merged, and the strong nuclear force appeared. In our era, when the average temperature of the universe is barely above zero K, the electromagnetic and weak nuclear forces appear separate, although the theoretical link remains.

The fundamental forces in the universe remain distinct, but the dream of unification of the sciences remains. Let's imagine that physicists finally succeed in combining the present three forces into a single comprehensive theory. Would that complete the consolidation of the sciences and become the "theory of everything?" What difference would that make for the physiologist deciphering the operating principles of the kidney? Or what difference would the unified physics theory make to the environmentalist or the economist? Very little, I suspect. While it would be exciting and reassuring to have a single theory describing the basis of all matter and fundamental interactions in the universe, uniting the sciences requires viewing them along another axis entirely: the axis of complexity.

UNIFYING THE SCIENCES THROUGH COMPLEXITY

It is common knowledge that you and I and everything we see and touch are all made up of fewer than 100 types of atoms. Without any special expertise in science, everyone also knows that chemists' or biochemists' molecules are made up of physicists' atoms,

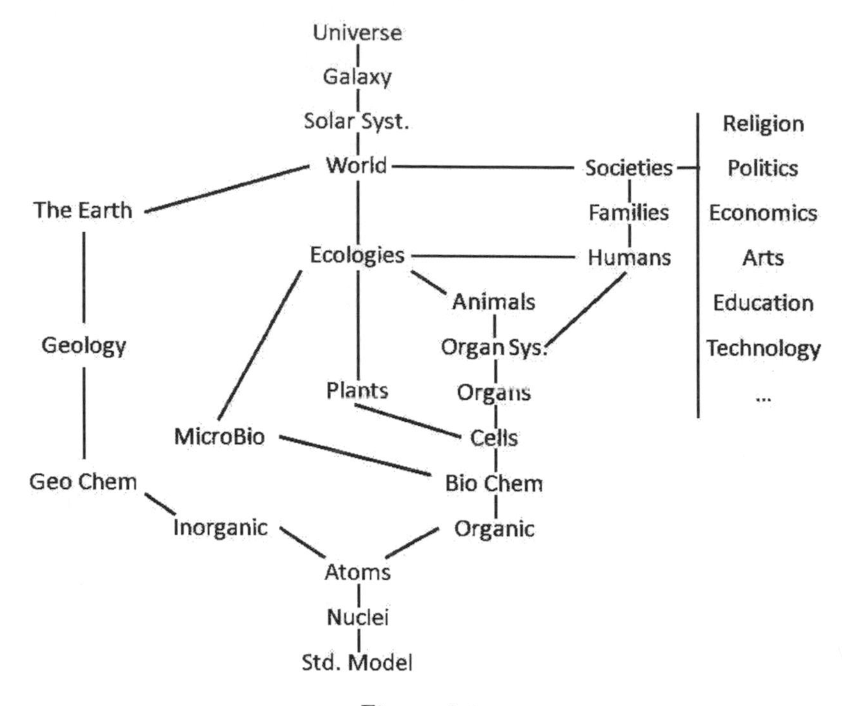

Figure 2.1
A simplified Catalog of the Universe

biologists' cells are made up of biochemists' molecules, physiologists' organs are made up of biologists' cells, etc. This relationship series is pictured as a hierarchy in Figure 2.1. Note that the structure has been extended upwards on the life sciences side through humans and their groups, on the left side to include inorganic materials and ecologies, upward to the cosmos, and downward below Atoms to the sub-nuclear areas. There are branches for specific inorganic materials and different forms of life.

The Sears, Roebuck and Company catalog played an essential role in American life beginning in the late 19th century. [3,4] Because of the 1862 Homestead Act and railroads' growth, a sizeable dispersed population existed in the Midwest and West – people with no ready access to home supplies, ready-made clothing, and tools. Richard Sears sold watches and jewelry, and in 1888 he published a mailer for his wares under the title "R.W. Sears Co.," promising that he would guarantee any watch sold to work accurately for at least six years. This advertising was so successful that his 1894 mailer offered a greatly expanded range of products, including sewing machines, sporting goods, musical instruments, saddles, firearms, buggies, bicycles, baby carriages, and men's and children's clothing. Each succeeding year, the catalog offerings grew in number and variety. Sears was the largest retailer in the United States until Walmart and Kmart surpassed Sears in 1990. The Sears yearly publication essentially defined the meaning of "catalog" in the minds of Americans. Children and adults alike spent countless hours through the year, pouring through its pages and dreaming about a better life with all those enticing products. Even when purchases were few, the Sears catalog defined what was possible. Sears stopped publication of the big catalog in 1993 but continues to send out various smaller specialized catalogs. In the 21st century, the Internet and search engines have replaced the comprehensive printed catalog.

I call the structure in Figure 2.1 the "Catalog of the Universe" because, in its complete and theoretical form (not the highly abbreviated form in the figure), this hierarchy defines what physical objects are possible in our universe according to current understanding. This Catalog does not take orders, of course, but it has an essential role in understanding the sciences and human activities, as I will show in the following chapters. Numerous precursors have preceded the Catalog, beginning with Aristotle's *scala naturae* and including the trees of life of the 19th century. The Catalog is unique in adding the inorganic and materials branches and the groupings of objects such as ecologies and cooperating human groups. Figure 2.1 includes a list on the right attached to (Human) Societies, naming some of the many specialized relationships humans have developed to achieve different goals. This list also represents Western higher education structure quite accurately since most university departments grew up around research areas. One glaring exception is mathematics, and later I will discuss the vital role mathematics plays in the origin of the entire Catalog structure.

The investigation of the characteristics of the Catalog of the Universe is the primary purpose of this book. The new worldview based on science and represented by the Catalog began to take shape in the 18th and 19th centuries as isolated research fields and associated individual academic disciplines. This Catalog is, in effect, the scientific successor to the Chain of Being that dominated thought for nearly 2000 years. The Catalog enables us to arrive at productive linkages between all sciences and humanities.

No God symbol tops the Catalog of the Universe, as in the Great

Chain of Being, and it contains no supernatural branch. The reason is that this Catalog includes only entities that can be observed and studied by science in the actual physical universe. It does not mean that the Catalog is an atheist concept. The realm of the supernatural – if it exists – is not included in this hierarchy because that realm is not identifiable in the natural world. I will have more to say about God and the supernatural in a later chapter.

The Catalog does not represent the universe's structure. You can't see parts of the Catalog through a microscope or a telescope because it is not a physical object. It has a tree-like structure, but it isn't a physical tree. Instead, it is an *organized relationship* between all the existing and possible objects in the universe: hence the name "Catalog." Suppose you wanted to construct a new, functioning universe based on the same laws that govern our universe. In that case, you could select a wide range of objects, living and inorganic, from the Catalog to populate your new universe. As we will see in Chapter 16, an appropriate structure for a universe is a network or a hierarchy of networks.

In recent years, cosmologists have begun to suspect the existence in the universe of two kinds of entities not made of atoms: dark matter and dark energy. No one has yet observed these entities directly because they do not emit or reflect light, but their presence is inferred from theoretical models of the universe's structure. I have not included either of the dark entities in the Catalog of the Universe because their existence and properties are not certain. If they exist, their relationship to atom-based matter is unknown.

I believe the Catalog's overall structure and its implications are not widely appreciated, either by scientists or by the public. I hope to convince you that substantial benefit in understanding can result from studying the Catalog, for example, understanding the reasons for the unique methods used in different science areas. Why were the scientific benefits of linking disciplines -- multi-disciplinary science -- not realized until the latter part of the 20th century? Why has exceptionally rapid progress been made since that time, and where are the most severe present limitations in scientific understanding? Fields such as Religion and Politics appear in the Catalog as group activities. I will elaborate on this choice later.

In this chapter, we saw how the new science developed first as isolated fields. It would take a long time before the areas began to cooperate in a significant way. For three and a half centuries, the emphasis was on observing, recording, and organizing observations. In the following chapters, we will see more about why a Catalog of the Universe is an appropriate structure to organize scientific thinking and much more in the 21st century.

CHAPTER 3

What Is a Catalog of the Universe?

> It is the harmony of the diverse parts, their symmetry, their happy balance; in a word it is all that introduces order, all that gives unity, that permits us to see clearly and to comprehend at once both the ensemble and the details. *Henri Poincare.*

Chapter 2 proposed an organizational structure that links all existing and possible objects in the universe: a tree-like diagram. There is a physical reason why this organization is appropriate for relating the sciences. We can arrive at that reason by starting with a question: What scale measures the vertical height of a subject area in this Catalog of the Universe? On the Chain of Being, height indicated a level of "Soul" with God at the top being 100% soul and rocks at the bottom with 0% soul. "Soul" is not going to be a suitable scale to measure height in this scientific structure! For the Catalog in Figure 2.1, two different and related scientific scales are possible.

Atoms are the constituents of all matter. Further, all living things are either cells or made up of cells, most of which are very small. The philosophical idea of a basic unit of matter is old. The ancient Greeks coined the word "atom." Aristotle postulated that earth, water, fire, and air are the ingredients that make up all matter. Other cultures, including Egypt, Babylonia, China, Japan, Tibet, and India, had similar elements. Medieval alchemists added to this list sulfur for combustibility and mercury for metallic properties. The ingredients were primarily philosophical, however, and not based on physical observations

It was not until the 19th century that the idea of the "atom" as a basic unit of matter had credibility as a scientific concept. Early in the century, John Dalton used atoms to explain why elements always react in ratios of small whole numbers. By the end of the century, scientists could estimate an atom's approximate size from experimental observations. The book *Les Atomes* published by Jean Perrin in 1913, cataloged 16 different measurements that made it possible to estimate the number of atoms in a gram atomic weight of an element or the number of molecules in a gram molecular weight of a compound, a number called *Avogadro's number*. The measurements included radioactivity, gasses viscosity, Brownian movement, and sky blueness. Also, at the end of the 19th century, J. J. Thompson discovered the electron and proposed a structure for the atom: a positive cloud of charge with the electrons embedded in it, the model given the name "plum pudding." His proposed structure was wrong, however, and it required the discovery of the nucleus early in the 20th century by Hans Geiger and Ernest Marsden and ultimately the development of quantum mechanics to arrive at more accurate structures for atoms.

Biological cells were first observed in 1665 by Robert Hooke. He did not see living cells or anything of their internal structure but rather the hollow cavities left by cells in cork as the material dried out. Antonie van Leeuwenhoek built a higher magnification microscope and reported to The Royal Society in 1676 observation of small moving objects he called "animalcules." It was years before anyone recognized the biological significance of cells. In 1839, Theodor Schwann and Matthias Jakob Schleiden proposed three tenets of *cell theory*. Two of them are still accepted:

1. All living organisms are composed of one or more cells
2. The cell is the most basic unit of life.

Schwann and Schleiden's third tenet proposed that a crystallization process formed cells. Within a few years, this tenet was shown to be incorrect, and it was replaced by

3. All cells arise only from pre-existing cells.

There are two physical reasons why a biological cell must be small, both reasons having to do with material exchange between cell interior and exterior. [1] Cells utilize energy to stay alive and accomplish their functions, and thus they need to move molecules in and out through their surface membrane and move molecules around internally.

To illustrate what would happen with large cell sizes, think of a spherical cell. A sphere's volume varies as its radius cubed, while its surface area varies as the square of its radius. As the radius increases, the volume grows faster than the surface area. In other words, as a sphere grows in diameter, there is less surface area for each unit of internal volume. For a cell, this would mean less ability to move energy sources in and waste out. Biologists concluded that above a radius of about 200 microns (0.008 inches or about twice the diameter of a human hair), a cell could not move enough molecules in and out through its surface to sustain life.

The second physical reason for limited cell size concerns the mechanism that moves molecules from place to place inside or outside a cell. That mechanism is diffusion, the tendency for molecules to move from regions of high to low concentration. The time for a molecule to move by diffusion a distance x increases as the square of x. For example, it takes four times longer to travel twice a given distance. Diffusion is an efficient and fast molecular transport method in water for distances up to about 100 microns, but it is very slow above this distance. Animals and plants, therefore, employ fluid flow, as in blood vessels, for molecular transport over distances above 100 microns.

Biological cells are not always spheres, of course. A cell can be long, provided that its diameter is small to meet the surface-to-volume and diffusion requirements listed above. For example, skeletal muscle cells range from 3 cm (1.2 inches) to 30 cm (12 inches) in length, and some nerve cells are even longer.

Because of the understanding of atoms and cells available since the middle of the 20th century, it is now logical to correlate size and complexity for both living and non-living materials. We could choose either one as a vertical scale for Figure 2.1. All ordinary matter in the known universe is made up of atoms so that larger things have more of them with more interactions between them, with cells as a higher basic structure for living matter. I have chosen "complexity" as a vertical scale for the Catalog since that characteristic is more challenging to understand than size or the number of atoms or cells.

There is a structural difference between one pound of sand and a one-pound animal, even if they have approximately the same number of atoms. The quantities *order* and *information* describe this difference. The animal represents high order and high information (describing its biological structure). The sand represents low order and high information (to describe the location of all atoms or sand grains) because these locations are not very meaningful. Thus, in the Catalog of the Universe, the inorganic left branch represents objects or material of less order than objects of the same approximate size in the organic branch.

The Catalog of the Universe has a structure described by a simple rule. The qualities of any object or material at *any* node in the Catalog are determined entirely by

1. the number and characteristics of the elements from the neighboring lower level(s) and
2. the nature of the relationships between the elements making up the object or material.

This rule applies everywhere in the Catalog, inorganic or organic, from nuclei to the cosmos. Types of molecules differ because they contain different numbers and types of atoms. But molecules with the same numbers and types of atoms can still have other properties when the relationships between their atoms are different. If you know an object's constituents that is insufficient to characterize it, you must also know the constituents' relationships.

To make this very personal, you – the reader – are made up of large numbers of electrons, protons, and neutrons. If someone disassembled you into three swarms of like particles, your identity and qualities would be entirely gone because all electrons, for example, are identical in every way and indistinguishable, having no memory of their history. [2,3] No one would ever know they had been part of you. Similarly, if you take a building apart brick by brick and stack them up without mortar, the building is entirely gone. It cannot be reconstructed from the brick pile without a knowledge of all the former brick-to-brick relationships.

You might be wondering why religion, education, the arts, and many other human activities also appear in the Catalog, along with the fields we usually classify as "sciences." Two qualities can also describe human social relationships: the number of individuals interacting and the Nature of their relationships. For that reason, all kinds of human activities appear in the Catalog of the Universe, from families (biological and marital relationships) to the myriad of social interactions humans participate in, from scientific research to education, entertainment, technology, philosophy, and religion. The list on the right-hand side of Figure 2.1 is illustrative, not complete. Later chapters will be devoted to human activities and how they compare with other levels in the Catalog.

Height in the Catalog in Figure 2.1 is thus a measure of complexity. Everything we know about, from an amoeba to a galaxy and including humans, is made of atoms. A larger size means more atoms relating in more complex ways.

In the next chapter, I recognize the Catalog of the Universe as a hierarchy and consider its qualities.

CHAPTER 4

The Catalog is a Hierarchy

> The hierarchy of relations, from the molecular structure of carbon to the equilibrium of the species and ecological whole, will perhaps be the leading idea of the future. *Joseph Needham.*

This chapter will show that the Catalog of the Universe is a hierarchy, describing what gives it that quality and why that structure is significant. Chapter 16 will address a contrasting type of organizational structure -- the network – and compare networks to the hierarchical Catalog.

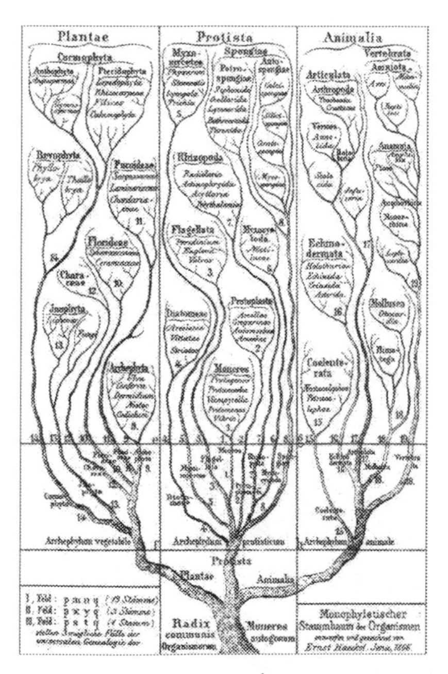

Figure 4.1

Ernst Haekel's Tree of Organisms, 1866[1]

In the Western world, hierarchies began in religious organizations. According to Wikipedia, the English word "hierarchy" was probably first used in the 1880 Oxford English Dictionary to refer to celestial and ecclesiastical hierarchies named Pseudo-Dionysius Areopagite (5-6 century CE). The Roman Catholic and Eastern Orthodox churches had hierarchical organizations, and the term later came to refer to similar organizational patterns in secular settings, such as management hierarchies. In general, hierarchies entered modern science in the biological taxonomy of Carl Linnaeus as described in his *Systema Naturae* (1735) and following publications.

All hierarchies include multiple levels, but the structures can serve different purposes according to the objects' relationships at different levels. The best-known type of hierarchy in the 21st century is the command hierarchy found in the military, businesses, churches, and other large organizations. In this type, plans or orders are generated at top levels and carried out at lower levels.

An *aggregational* hierarchy organizes information for convenience and clarity. [2] Examples of this type are the Scala Naturae of Aristotle, the Great Chain of Being described earlier, Linnaeus' taxonomy in biology, and the many computer databases accumulated by today's computers. Figure 4.1 is an example of a "Tree of Life" hierarchy from the 19th century designed to organize biological species.

The hierarchy type of interest in this book is the *constitutive* or *structural* hierarchy where objects at any level consist of linked groups of things from the next lower level. The Catalog of the Universe introduced in Chapter 2 is such a structural hierarchy.

Francis Bacon (1561-1626) defined the scientific method, a repeating experiment and theory sequence that replaced natural philosophy. Bacon also described an organization of knowledge at the beginning of the 17th century. [3] He started with three fundamental divisions: history, poetry, and philosophy and subdivided each of these into three aspects: divine, human, and natural. The result was nine divisions in the form of a branching tree.

The first person in the Scientific Era organizing the sciences in a hierarchy was the 19th-century French philosopher Auguste Comte (1798-1857). Comte is best known as the founder of sociology and somewhat less as the originator of a philosophy called Positivism. But he was also interested in the relationship between the sciences. He came up with the order (least complex first): Mathematics, Astronomy, Physics, Chemistry, Biology, and Sociology. Here is how Comte described his hierarchy:

> "It is reserved for our own century to take in the whole scope of science; but the commencement of these preparatory studies dates from the first astronomical discoveries of antiquity. Natural Philosophy was completed by the modern science of Biology, of which the ancients possessed nothing but a few statical principles. The dependence of biological conditions upon astronomical is very certain . . . One connecting link was supplied by the science of Chemistry which arose in the Middle Ages. The natural succession of Astronomy, Chemistry, and Biology leading gradually up to the final science, Sociology, made it possible to conceive more or less imperfectly of an intellectual synthesis . . . In the seventeenth century, however, the science of Physics specially so called, was founded; and a satisfactory arrangement of scientific conceptions began to be formed. Physics included a series of inorganic researches, the more general branch of which bordered on Astronomy, the more special on Chemistry. To complete our view of the scientific hierarchy we have now only to go back to its origin, Mathematics; . . . We see from these brief remarks that the series of the abstract sciences naturally arranges itself according to the decrease in generality and the increase in complication. We see the reason for the introduction of each member of the series, and the mutual connexion between them . . . That co-ordination at once establishes unity in our intellectual operations. It realizes the desire obscurely expressed by Bacon for a *scala intellectûs*, a ladder of the understanding, by the aid of which our thoughts may pass with ease from the lowest subjects to the highest, or vice versa, without weakening the sense of their continuous connexion in Nature." [4]

Several other later authors also proposed incomplete Nature-based hierarchies based on concepts other than complexity (for a summary, see Salthe [5]). In the early 20th Century, J. Needham recognized that a hierarchy was going to be necessary to understand mechanisms in biology:

> "Whatever the Nature of organizing relations may be, they form the central problem of biology, and biology will be fruitful in the future only if this is recognized. The hierarchy of relations, from the molecular structure of carbon compounds to the equilibrium of species and ecological wholes, will perhaps be the leading idea of the future." [6]

J. H. Woodger was an early theoretical biologist, and he wrote, also in the early 20th century

> "But from what has been said about organization it seems perfectly plain that an entity having the hierarchical type of organization such as we find in the organism requires investigation at all levels, and investigation of one level cannot replace the necessity for investigations of levels higher up in the hierarchy." [7]

Arthur Koestler, in *The Ghost in the Machine,* wrote of many types of hierarchies [8]. There are social hierarchies of family and clan and geographical distribution. There are hierarchies in living organisms' functions, from instinctive behaviors to sophisticated skills like talking and piano playing. The processes of becoming are hierarchical: phylogeny, ontogeny, and the acquisition of knowledge.

It is not my purpose in this book to address the origin of the objects presently populating the Catalog of the Universe. William C. Burger has recently described the basis of the Catalog's biological portion in his book *Complexity: The Evolution of Earth's Biodiversity and the Future of Humanity.* [9] The goal of the present book is to draw some conclusions from the entire hierarchical structure about the advance and challenges of scientific research and human group activities.

Is the scientific Catalog an essential feature of the universe, or simply a pattern generated by the human brain to organize the sciences? Koestler quoted a parable from Herbert A. Simon of early computing fame, illustrating the importance of hierarchical assembly for the origin of biological systems:

> "There were once two Swiss watchmakers named Bios and Mekhos, who made very fine and expensive watches. Their names may sound a little strange, but their fathers had a smattering of Greek and were fond of riddles. Although their watches were in equal demand, Bios prospered, while Mekhos just struggled along; in the end he had to close his shop and take a job as a mechanic with Bios. The people in the town argued for a long time over the reasons for this development and each had a different theory to offer, until the true explanation leaked out and proved to be both simple and surprising.
>
> The watches they made consisted of about one thousand parts each, but the two rivals had used different methods to put them together. Mekhos had assembled his watches bit by bit rather like making a mosaic floor out of small colored stones. Thus each time when he was disturbed in his work and had to put down a partly assembled watch, it fell to pieces and he had to start again from scratch.
>
> Bios, on the other hand, had designed a method of making watches by constructing, for a start, sub-assemblies of about ten components, each of which held together as an independent unit. Ten of these sub-assemblies could then be fitted together into a sub-system of a higher order; and ten of these sub-systems constituted the whole watch ....
>
> Now it is easy to show mathematically that if a watch consists of a thousand bits, and if some disturbance occurs at an average of once in every hundred assembling operations -- then Mekhos will take four thousand times

> longer to assemble a watch than Bios. Instead of a single day, it will take him eleven years. And if for mechanical bits, we substitute amino acids, protein molecules, organelles, and so on, the ratio between time-scales becomes astronomical; some calculations indicate that the whole life-time of the earth would be insufficient for producing even an amoeba -- unless he [Mekhos] becomes converted to Bios' method and proceeds hierarchically, from simple subassemblies to more complex ones."

Simon then went on to say:

> "Complex systems will evolve from simple systems much more rapidly if there are stable intermediate forms than if there are not. The resulting complex forms in the former case will be hierarchic. We have only to turn the argument around to explain the observed predominance of hierarchies among the complex systems Nature presents to us. Among possible complex forms, hierarchies are the ones that have the time to evolve." [10,11]

The Catalog of the Universe is thus more than just an organizational form to aid human understanding. Rather, it is an important result of natural processes and essential for evolution. Recent proposals picture the brain as a hierarchy [12] and that all languages are fundamentally hierarchies. [13]

Koestler invented the new name "holon" for an object at any level in a structural hierarchy, along with "holarchy" for that type of hierarchy. [8,14,15] Why are the new names needed? Because an object at any level can have two personalities. It can be a self-contained object, a "whole" with *agency*. It can also be a "part" interacting with others at the same level -- a property of *communion*. Holons at any level in a hierarchy thus have a two-faced Janus-like character. For example, biological cells are complete entities and can live independently in the proper environment (temperature, chemical constituents, etc.). But they can also interact with other cells. Holons also have two other properties in the vertical direction of the hierarchy. If a holon does not maintain its agency and communion properties, it can decompose into sub-holons, a process called *self-dissolution*. Linking with other holons is the opposite, a creative event from which new holons emerge at a higher level. This process is called *self-transcendence* or *emergence*. The names "holon" and "holarchy" were coined to emphasize this type of hierarchy's unique qualities.

The number of different kinds of holons at any level in a hierarchy is called its "span." For example, the molecular level span is enormous, while the span at the biological organ systems level is relatively small. The number of levels in a hierarchy is its "depth."

A hierarchy can have a bottom (lowest level) and a top (highest level). The Catalog of the Universe has neither of these. Noone knows or even suspects that the Standard Model of nuclear physics is the lowest subdivision of matter in our universe. Similarly, there is no clear highest level, either of human, ecological, or cosmic organization. Such a hierarchy is called an Open Hierarchical System.

Koestler described ten characteristics of Open Hierarchical Systems in his Appendix I. He considered his list to contain elements from the trivial to those that need more confirmation. Paraphrasing Koestler's characteristics briefly, they are:

1. The **Janus Effect**, i.e., holons at any level are both compound – made up of holons at a lower level in the hierarchy – and possible components for holons at a higher level.
2. Hierarchies are **dissectable,** i.e., organized as levels linked by relationships and communications.
3. Functional holons are governed by fixed sets of **rules** and display more or less flexible **strategies**.
4. Every holon has the dual tendency to preserve and assert its individuality as a **Semi-Autonomous whole** and function as an **Integrated part of a larger whole**.
5. Output hierarchies generally operate on the **trigger-release principle**, where a relatively simple signal releases complex, preset mechanisms.

6. Hierarchies can be 'vertically' **arborizing** structures whose branches interlock with those of other sub-hierarchies at multiple levels and form 'horizontal' networks (**reticulation**).
7. **Regulation** in a hierarchy usually proceeds down through the hierarchy, not through bypassing channels around levels of the hierarchy.
8. Holons on **higher levels** show increasingly complex, more flexible, and **less predictable** activity patterns, while those on **lower levels** are increasingly mechanized, **stereotyped**, and predictable.
9. A holon or organism is said to be in **dynamic equilibrium** if its sub-holons' Self-Assertive and Integrative tendencies counterbalance each other. **Disorder** is the state where the sub-holons are out of control.
10. Critical challenges to a high-level holon (organism) can produce generative or **regenerative** effects.

The Catalog of the Universe is not a description of the existing world and universe. Instead, the Catalog levels are *categories* of things (holons) of different complexity that can or do exist. A hierarchy populated with holons at every level is like a catalog, from which many kinds of objects could be selected and allowed to interact to form a real universe. The Catalog is a big one since the biological branch alone contains over a million observed species. Biologists estimate that the actual living number on earth is about 10 million, and the number that lived on earth since its formation is a hundred times larger! A different organizational structure is needed to describe that real universe of interacting holons: a network. Chapter 16 of this book will describe networks.

To review, the relationship between **categories** of holons of different degrees of complexity is a hierarchy of holons or a holarchy. In the next chapter, I will discuss a universal principle that operates at all levels of the holarchy, the Catalog of the Universe.

CHAPTER 5

Relationships Create the Catalog

> What is this mind of ours: what are these atoms with consciousness? Last week's potatoes! *Richard Feynman*

The holons at any level in the Catalog of the Universe can differ in two possible ways: their numbers of sub-holon types and the interactions between their sub-holons. Sub-holons are like bricks; their interactions or relationships are the mortar. The Catalog exists because of relationships. In this chapter, I consider the universal importance of relationships.

Relationship is not a new concept in science. Aristotle wrote about connections between living things, but his relationships meant relative positions in the Chain of Being. [1] This was as far as anyone could go for many centuries in natural relationships until scientists recorded more observations, and theorizing changed from deductive to inductive.

George Henry Lewes expressed early recognition of the importance of relationships in science in an 1879 volume:

> "No fact is explained by the enumeration or exhibition of its factors as *isolated* elements; only by these in their combination and mutual dependence. Comte was guilty of an oversight when he defined the chemist's problem to be that of 'determining the properties of compounds by the properties of their components,' for this is impossible. The properties of water could never be determined by enumerating the properties of oxygen and hydrogen; no salt is discernible in its acid, nor in its base. The properties of compounds must be observed in the reactions of the compounds." [2]

A well-known example of the scientific importance of relationships is Einstein's relativity theory. Here is the Special Relativity version: "The laws of physics are the same for two observers moving, but not accelerating, with respect to each other, and the speed of light (in a vacuum) is independent of the motion of the observers." The exception for acceleration is critical and required Einstein to continue work to develop General Relativity. For this discussion, the important point is that physics laws change for objects depending on their relative and changing locations. How did Einstein come to recognize the importance of relationships? There is some evidence of influence from an older scientist. [3]

At age 17, Einstein became friends with 23-year-old Michele Besso, and the two remained close friends for many years. Besso was brilliant and made an ideal sounding board for Einstein's ideas, although he was somewhat scatterbrained in other areas of his life. One weekend, Besso presented Einstein with two books by Ernst Mach -- the Mach who gave his name to a measurement of speed relative to sound speed, like Mach 1, Mach 2, etc. After that, the two discussed Mach's ideas frequently.

Mach was a dualist early in his life, believing that mind was not a physical brain property but was rather a non-physical entity like soul. He was a keen observer of the natural world. At age 15, he suddenly realized that all the sensations experienced in his observations were physical effects impinging on his physical brain. The view he finally arrived at is now called *neutral monism* and describes every aspect of reality, physical objects and sensations, as purely relational, element to element. Even a measurement is a relation. For example, length is a relation between an object and a ruler. Here are Mach's words:

> "The aim of research is the discovery of the equations which subsist between the elements of phenomena. The equation of an ellipse expresses the universal conceivable relation between its co-ordinates, of which only the real values have geometrical significance. Similarly, the equations between the elements of phenomena express a universal, mathematically conceivable relation." [4]

Einstein had been struggling with the knowledge that no one had been able to devise an experiment to determine their motion as an observer relative to absolute space. A scientist could measure only movement relative to some other object. If the speed of light is 186,000 miles per second for the observer and the observer is moving with respect to an object, shouldn't the light speed from the object's viewpoint, just like a sound changes its pitch when the source is moving toward or away from the listener? Einstein traveled to see Besso, and they hammered on this problem at length. The next day, Einstein returned to Besso, thanked him for his interactions, and explained his solution. As unphysical as it seems at first thought, the speed of light (in a vacuum) is always the same regardless of the relative motion of source and observer. Within a few weeks, the theory of Special Relativity was complete. Einstein sent his paper to Mach and, to his great joy, received the reply, "Your friendly letter gave me enormous pleasure."

Another relationship discovered by Einstein is more famous than that of objects in relative motion: $E = mc^2$, expressing the idea that mass can be converted into energy and vice versa. All processes in the universe, from the lives of bacteria to galaxies, can be described as interconversions of different forms of mass and energy. The energy radiated by the sun, for example, comes from the conversion of a tiny fraction of the mass of pairs of hydrogen atoms into helium atoms. When the sun's radiated energy reaches the earth, some gets converted by plants into more plant mass. You and I consume the plants and utilize some of their stored energy to power all our life processes – including our brain and thinking.

Einstein realized that the heart of reality was not the objects or material in the universe but rather their relationships.[5] As expressed by physicist Richard Feynman in the quote at the beginning of this chapter, last week's potatoes get converted to brain (and other) cells! The relationships that attracted Einstein's attention were symmetries, situations that remain unchanged when transformed. For example, if the two people on a balanced teeter-totter change places, the balance remains. A perfect sphere remains the same when rotated about any axis. Following up on Einstein's focus on symmetries, the German mathematician Emmy Noether was able to show that each type of symmetry Einstein and other physicists noticed is associated with a conservation law. Examples include the conservation of energy, electric charge, and momentum. A recent example of applying symmetry to advance scientific understanding is predicting [6] and later experimentally discovering the Higgs boson.

The Catalog of the Universe is an expression of the vital role of relationships in the universe. Atoms join to create molecules, molecules combine in a system to make a cell, and cells link up to form plants and animal organs on up to humans cooperating to create music or business and stars clustering to form galaxies. The immense variety of holons at all these levels results from the characteristics of the holons coming together (the bricks) and the qualities of their relationships (the mortar).

Types of relationships in the Catalog vary with the level, and the relationship complexity also varies from level to level. For example, atoms can interact in only a few ways to create molecules. Higher in the Catalog, individual humans can interact in an immense variety of ways.

From human, animal, and plant physiology, there is the cooperative action of large numbers of similar cells to form macroscopic organs. Examples are the brain, skeletal muscles, the heart, and smooth muscle in blood vessels and intestinal walls. Each of these example organs is a linked community of cells of similar types. Each organ's functional operation is due to its constituent cells' properties, geometrical relationships, and their electrical, chemical, and mechanical communications. In writing his comprehensive history of biological thought, Ernst Mayr described how relationships were always on his mind in his biological research:

> "Perhaps the most important aspect of holism is that it emphasizes relationships. I myself have always felt that relationships are not given sufficient weight. This is why I have called the species concept a relational concept and why my work on genetic revolution (1954) and on the cohesion of the genotype (1975) both deal with relational phenomena." [7]

Biologists are familiar with interactions between different types of organisms. *Competition* is one of the best-known types of biological relationships because of its role in the theory of evolution developed by Charles Darwin and Alfred Russel Wallace and published by Darwin in *On the Origin of Species* in 1859.

Symbiosis is the opposite of competition. The word, earlier used for human communities, was first applied to lichens in 1877 by Albert Bernhard Frank. There are cooperative communities in biology where individuals share functions necessary for the community's survival, including nests of ants and bees and African termites that construct large mud residences. Even trees of different species communicate and share resources and defenses through their roots linked together by fungi networks. [8,9] Another apparent symbiosis is the synchronized flashing of large groups of fireflies, although their cooperation benefits are not well understood. [10]

Commensalism is a relationship where one species derives benefit from another but does not hurt or benefit the other. The opposite is *parasitism,* where the parasite harms the host. In the *predator-prey* relationship, the predator seeks out and destroys the prey.

The ultimate natural relationship is an ecological network of multiple life species, from single cells through plants and animals (including humans), interacting with each other and local geology, weather, and seasons. Chapter 16 will consider ecological networks in more detail.

An early description of interactions in nature was written by J. Arthur Thomson long before he or others were able to study multiple relationships in detail:

> "It is part of this order that the world is a network of interrelations. Part is linked to part by sure, though often subtle bonds, and nude isolation is as rare in nature as a vacuum. Nature is a vast system of linkages. . . . Only the working naturalist knows the extent to which living creatures are interlinked in nature. There is a solidarity of kinship, but there is also a solidarity of vital relations. We are familiar with the correlation of organs in the living body, but there is also a correlation of organisms in the web of life." [11]

Thus, relationships are the glue holding holons together throughout the Catalog, making an enormous range of objects possible in the universe. I will have more to say about relationships high in the Catalog later in this book. Meanwhile, Chapter 6 will describe scientists' methods for gathering the information and ideas that give the Catalog its structure.

CHAPTER 6

How Is Scientific Research Done?

> Science is the enterprise that prods reality to answer us back when we're getting it wrong . . . Science is, by definition, the methodology that enlists reality itself as collaborator, and what methodology could possibly compete with so successful a collaboration? *Rebecca N. Goldstein*

In the previous chapter, we learned that many object types exist in the universe because of relationships. I can organize a Catalog of object types because of many scientists' work, curious individuals eager to find patterns and regularities in some aspect of our universe. Each of these scientists or groups typically makes contributions to understanding at one or two levels. The Catalog is a roadmap that shows how all the individual fields of concentration are related. While scientists utilize many different means of experiment and observation, there is a very definite methodology in how they all go about their search for understanding. This chapter describes that methodology.

As modern science had its beginnings in the 16th and 17th centuries, the first research emphasis was on collecting and documenting observations and facts. Scientific treatises of these and even those of the following two centuries often contained hundreds of pages of detailed observations. Starting in the 17th century and growing after that, theories were gaining prominence in the physical sciences. It took another century or more before theorizing began in a serious way for biological and medical sciences. As Sir Lawrence Bragg, Nobel Prize winner with his father for analyzing crystal structures with x-rays, wrote in *The History of Science*, the essence of science "lies not in discovering facts, but in discovering new ways of thinking about them."

In the Prologue of this book, I described my early pastime of taking apart different things around the house, like toasters and radios. I had a natural interest in how various devices worked: what mechanics, electricity, and electronics principles made the device function? Books and *Popular Science* magazines provided insights along with suggestions for experiments. My learning led to a series of jobs in repair shops that were much more interesting than the usual teenage lawn mowing jobs. I did not realize that I was following the same steps to discovery pursued in scientific research laboratories.

Looking for understanding, scientists generally use the "taking-apart" method, formally called *reduction*, in constructing theories for the systems they study. A scientist will dissect a system into parts and the map of their interconnections, a process equivalent to looking down the Catalog of the Universe one level for understanding. For example, in some biological and medical studies, the dissection can be physical – a dissection into organs, cells, or subcellular organelles. In other cases, the dissection must be theoretical: guessing what the internal components might be and how they might be interconnected. In either case, for the results to become a working theory, the component behaviors must also be known or assumed.

In the philosophy of science, reduction refers to a succession of theories, subdividing objects at each level to produce a new lower level. *Ontological reductionism* is the belief that any phenomenon can, theoretically, be understood entirely and the ultimate truth attained by a reductive succession of theories, each one depending on a more basic theory until a fundamental elementary level is achieved. *Methodological reductionism* is also the practice of seeking understanding by a reductive succession of theories. But in this case, there is no requirement that the hierarchy of theories must extend down to a fundamental elementary level. Today we know that atoms are not a foundational level for the natural hierarchy, and neither are the protons, neutrons, and electrons that form atoms. We do not have a successful theory that could be considered an ontological foundation for the natural hierarchy. It is quite acceptable that the known hierarchy be open at the bottom as well as the top.

Modern science thus operates under the philosophy of methodological reductionism. This philosophy does not promise to deliver the "ultimate truth." Instead, the emphasis is on a practical understanding of the world and universe that can increase human survival and quality of life, satisfy curiosity, produce reasonable histories of the universe and predictions for the physical future, and help address existential questions. Thus scientists working at any natural hierarchy level only need to dissect their subject down one or a few levels to achieve adequate understanding. For example, a physiologist studying the heart does not have to anchor a theoretical model on quarks and electrons. Thinking in terms of cell properties or even cell biochemical reactions (i.e., one or two levels lower in complexity) is not only adequate but necessary. A complex system's properties cannot be derived from the properties of that system's components alone because new properties and new effective laws arise in the synthesis process. [1] There is also a practical limitation. Even if a scientist wanted to construct a heart theory based on quarks and electrons, there is no known way of studying a theory containing enough elementary particles to make up a physical object such as a heart or even a single heart cell! Even today's largest supercomputers are orders of magnitude too small to study such a theory. Animal descriptions in terms of organ systems, organs, and cell types are good science. Physicists have developed working theories for astronomical numbers of identical particles such as gas molecules by considering their average behavior rather than individual particle movements. Such averaging approaches, however, are generally not applicable to biological systems.

The "methodological" adjective also includes the limitation of theories to natural mechanisms (excluding the supernatural). The rule is a practical one and does not imply either atheism or theism. The exclusion of the supernatural is practical because its introduction into scientific logic stops further investigation. I will use "reduction" as a shorthand for "scientific methodological reduction."

Aristotle is recognized as originating science based on observation, in contrast to Plato's base of pure reason, although the observations Aristotle made were limited. Aristotle used intuition and methods of philosophy to provide theories for his observations, for example, observations of physical object movement, a process now called *deduction*. [2] 21st century science theories are based on *induction* from observations and experiments. Aristotle understood the universe as being made up of different combinations of four fundamental elements: earth, water, air, and fire. All objects were believed to have a natural place and a tendency to move to that place, for example, falling for earthy solids and rising for fire. In Aristotle's cosmos, the earth was the center of the universe, with the moon, sun, and planets each made up of a fifth element and revolving around the earth in circles. The heavenly bodies were believed to have souls or supernatural intellects to guide them in their travels.

Aristotle had a cadre of students learning from him, but the idea of learning about the natural world from observations mostly died out for the next twenty centuries [3]. Aristotle's understanding of the universe was promoted in the West for those centuries by the Christian church, giving Aristotle's ideas stature as authoritative with little or no further development.

The reduction method based on observation and experiment now seems like a natural and obvious way to search for explanations. Still, philosophers did not consider it until the early 20th century. Here is a definition of reduction by Nagel:

> "A reduction is effected when the experimental laws of the secondary [higher] . . . are shown to be the logical consequences of the theoretical assumptions . . . of the primary [lower] science." [4]

There is some evidence that reduction may even be a natural aspect of human brain function, either innate or learned, as a source of understanding. A sociological study by Hopkins and collaborators [5] showed that study subjects judge scientific explanations to be of higher quality when they contain information from the neighboring more reductive field, even when that information is not known to be relevant.

What did scientists at the middle or higher levels in the Hierarchy do before there was sufficient knowledge about lower levels to think about their systems in verifiable reductionistic terms, i.e., form theories for their system out of known simpler components? For example, what did biologists studying animals do before discovering organ systems like blood circulation or nervous systems? Or what did cell biologists do before there was knowledge about biomolecules? They did a great deal of observation, description, and classification and less theorizing for one thing. The

18th and 19th centuries' scientific books often contained many detailed collections of facts but little theory.

The only theoretical options for these early scientists were hypothetical lower-level components or philosophical arguments. For example, biological systems were assumed to be driven by a *vital force*, a principle distinct from chemical or physical forces. [6] From the ancient Greeks and Romans until the 19th century, physicians based their treatments of human illness on the theory that the body's health depends on a balance of four "humors": black bile, yellow bile, phlegm, and blood. Any illness could be explained by out-of-balance humors, i.e., too little or too much of any of these four. This theory was behind the medical treatment of bleeding ill patients, when, as we now know, those patients needed their blood more than ever!

Where did the humors theory come from? It is not known for sure. But suppose a container of blood is allowed to sit undisturbed for a while. In that case, it separates into a dark clot at the bottom ("black bile"), above it a layer of packed red blood cells ("blood"), followed by a layer of white blood cells ("phlegm"), and finally a layer of transparent yellow serum ("yellow bile"). It was well into the 20th century before scientists knew enough to abandon the "humors" concept and recognize body systems problems as the cause of illness.

In the early years of science, right up to the middle of the 20th century, individual researchers made the most advances. These pioneers may not have been thinking of their theorizing as reduction, but if they were to build their theories on knowledge of the adjacent lower level of complexity, as described in this book, working knowledge of the lower levels would be necessary. For example, a biologist trying to understand a nerve cell's function would need to understand biochemical reactions, membrane transport, osmotic pressure, and diffusion. Acquiring this breadth of knowledge would be increasingly challenging as each field developed.

The real solution to this dilemma was *teamwork* – research not done by isolated individuals but by multidisciplinary teams. I was fortunate in my postdoctoral work at Duke University to be part of such a multidisciplinary team, as described in the Prologue. At the time, I did not realize how unusual that was until a prominent scientist from another university spent a morning with our team in our usual energetic team discussion. As we left the room for lunch, the visitor said to me, "How do you stand that level of intensity every day and keep the team together?" With a multidisciplinary team, any system's assumed internal structure could be based on prior knowledge of the lower-level science rather than being hypothetical. The beginning of multidisciplinary teams in the mid-20th century substantially increased scientific research quality and productivity.

Reduction requires one more step after the dissection into components to make its results thoroughly convincing. That step is reassembling the components in their assumed relationships, demonstrating that the dissection was correct, that the components were undamaged in the dissection, and that their connections are accurate. Lewes expressed this idea in 1879:

> "Even in physical research the analysis which decomposes a total into several components must always be followed by a synthesis which reconstructs the whole, and thus, restoring all the suppressed conditions, reuniting what provisionally was separated, views the parts in the light reflected from the whole." [7]

Because reassembly or synthesis is challenging, it is often ignored – at the peril of continuing with an incorrect theory. With current technologies, physical reconstruction is difficult or impossible for many systems -- biological systems, for example. Reconstruction by mathematics is possible, however.

Mathematics plays a significant role in both constructing and gaining understanding from the Catalog of the Universe. In the following chapters, I will describe why mathematics is essential to science and why the advance of science could not have continued without computers.

CHAPTER 7

What is the Purpose of a Scientific Theory?

> What I cannot create, I do not understand. *Richard Feynman.*

There is a tool that applies to all sciences, although some sciences have benefited more than others from this tool. That tool is mathematics, and it is a tool for thinking. This chapter will explain why mathematics plays such an essential role in science, beginning with an early career experience.

After receiving a Ph.D. degree in physics and a 4-year postdoctoral opportunity to learn about challenging biomedical science problems, I joined the Cardiac Electrophysiology Laboratory in the School of Medicine of Duke University, as described in the Prologue. Heart cells are muscle cells, but a brief burst of electrical activity that is in some ways similar to the action potential of nerve cells triggers their contractions. One of the Duke research team's goals was to learn how heart cells generate their action potentials.

Not long after my arrival, the team learned about a relevant experiment conducted by researchers in Germany. About 20 years earlier, a research group in England deciphered the mechanism behind nerve cells' action potential by some very creative experiments on a nerve axon from the giant squid. (I will describe that project in Chapter 11.) Since the goal was similar for heart cells, the German group decided to use the same experimental design. But there was one problem. There are no giant heart cells (even from elephants or whales!), so the German group instead used a small bundle of heart cells. The bundle diameter was about the same diameter as the nerve axon in the earlier research.

The Duke research team thought the German experiment results were a little unusual, so the Duke team decided to create a detailed computer-based mathematical model of their experimental setup and tissue. I was assigned that task with guidance from the group leaders. It took only a few days to create and test the computer model, and the results were striking. They showed us that the German records were not measurements of the heart tissue sample, but instead measurements of the tissue bath used to keep the bundle of cells alive! The heart cell bundle may have had the same size and shape as the nerve axon, but it worked entirely differently in the experiment.

In the summer of 1971, there was a scientific conference in Munich, where we could present our cardiac electrophysiology research. This would be my first presentation at an international conference. The German researchers would be in the audience – listening to me describe how they had measured the bath rather than the tissue. The discussion following my presentation was memorable, to say the least. The next scheduled presenter was a no-show, so the session chair allocated even more time for the Duke versus Germany research discussion.

The lesson my colleagues and I learned from this experience was that intuition is an inadequate form of scientific theory. In this example, the sample tissue was the same size and shape as in the earlier nerve experiment. The interior structure, however, was very different. The use of mathematics, implemented in a computer, forced us to be very specific about experiment design details. In the long term, this interchange had a positive effect. Later electrophysiology researchers realized they must study and justify their experimental design with mathematics and not just publish recorded results.

Modern science aims to benefit human life by reducing mystery and uncertainty in human interaction with the environment and finding ways to adapt to the environment to increase human comfort and enjoyment. We are now past the period where the gods were assumed to control everything and the centuries where all believed that humans exist in a fixed hierarchy with no hope of change. In the age of science, we are learning to observe our environment and ourselves, and from those observations derive principles underlying what we observe and experience. That is the definition of *science*. The complementary activity is *engineering*: applying the discovered principles to increase human comfort and enjoyment and decrease human uncertainty about the future.

Science and engineering both depend strongly on human thinking, the ability to manipulate ideas and derive consequences. As I have described, science started with a few curious individuals, but it has developed into a social community of thinking where all interested persons can participate. For this community to function, a common thinking language is necessary; that language is mathematics. More than just a descriptive language, mathematics includes methods that can derive consequences from principles -- methods that can apply to *any* area of human thinking.

Mention "mathematics" in a conversation and, unless you are speaking with certain scientific colleagues, the chances are high that the response you get will be a shrug and a weak apology about fearing the subject. It is true that mathematics is abstract and that people vary in their ability to think abstractly. But, searching for the basic principles of the natural world without mathematics would be like expecting your mechanic to work without wrenches.

Gaining the understanding available from the Catalog of the Universe requires knowledge of a few essential mathematical concepts. I will attempt to present these concepts in the following chapters with a minimum of equations or symbols.

First, an explanation of the role mathematics plays in science. There has been a pattern in the way scientific fields developed at each level of the Catalog of the Universe, a series of steps based on the incremental increase of knowledge, producing organization first and finally understanding:

1. **Naming** – There is a Chinese saying that the first step to wisdom is getting things by their correct names. The first step in the birth of any scientific field is to observe and then name the objects or phenomena of interest. Fortunately, most areas recognized at the beginning that science needs a naming system.
2. **Grouping** – The next step in understanding is to group objects or phenomena with similar characteristics. For example, does the falling of different things have something in common? We realize

there is not just one hummingbird, but several with common characteristics like small size, rapid wing motion, and long beak, but differing in color.

3. **Relationships** – Additional organization levels follow, such as a hierarchy showing commonality levels between groups having some characteristics in common. For example, hummingbirds are part of a tree relating all types of birds. The bird tree (along with reptiles) then becomes a branch of a more comprehensive tree with animals, fish and other sea creatures, amphibians, etc. The process of developing the hierarchy of living things has produced many "trees of life" in biology over the centuries. Figure 7.1 shows the base of the presently-known Tree of Life with its three main branches: Eukaryotes (organisms made of cells with nuclei, from single-cell organisms to animals and plants), Bacteria (single cells with no nuclei), and Archaea (also single cells with no nuclei, but other characteristics different from Bacteria). You, the reader, and I, the writer, belong to a tiny branch, "*Homo sapiens,*" part of the "Animals" branch of Eukaryota in Figure 7.1.
4. **Theory: Physical Models** – The purpose of theory in science is to find fundamental principles or laws that account for the organized observations. Theory development usually starts with a drawing on a pad or whiteboard, a visual hypothesis showing possible known factors or natural laws that might account for the observations or a postulated new principle. An example might be a set of circles representing components in a system with lines between circles representing interactions between the components, creating an internal map of the system. The map becomes a model for thinking that hopefully has the same behavior as the observations. Since the model at this stage is purely an idea, its behavior must come from intuition, the internal voice that says: "I think this model will have the same behavior as our observations." Intuition plays a significant role in scientific research, such as constructing relationships in levels 2 and 3 above and designing experiments. While intuition can contribute to the initial development of theories, it can quickly go astray by itself and is a weak form of theory. Communities do successful science, and that requires a language to communicate

ideas accurately and precisely to colleagues. Intuition is personal and imprecise; only its results can be shared and not its logic. In some cases, a physical model can help. Examples are the balls and sticks used by chemists to represent molecules or the system of pump, tubes, chambers, and valves used to model blood circulation. Physical models are imperfect as theories because they are analogies, so their representations are always incomplete. Physical models are also limited when it comes to one necessary quality of a good theory: the ability to make predictions. Almost every physical model has some unwanted trait, such as friction, that can only be minimized but not eliminated.

5. **Theory: Mathematical and Computer Models** – Since intuition and imagination are unreliable for predicting a conceptual theory's behavior, what is needed is a tool that can mechanize concepts so they can interact with each other and produce the resulting sample behaviors. Mathematics, with its rigorous and well-defined methods, provides tools to build and test scientific theories. Expressing a theory in mathematics requires a precise and explicit description of intentions. No fudging is permitted, and nothing can be left unspecified. A mathematical theory can be simple or complex as needed to account for the observed data. Mathematical theories are repeatable and independent of human language or culture and thus ideal for sharing. Mathematics is a scientific field of its own, of course, and is opaque to many, even many researchers. Because mathematicians cannot solve some types of equations, there are limits to theories described and exercised in mathematics. Computers have made it possible to work with a much more comprehensive range of equation types. Some additional explanations follow in the next chapters because of the importance of mathematical theories to science development.

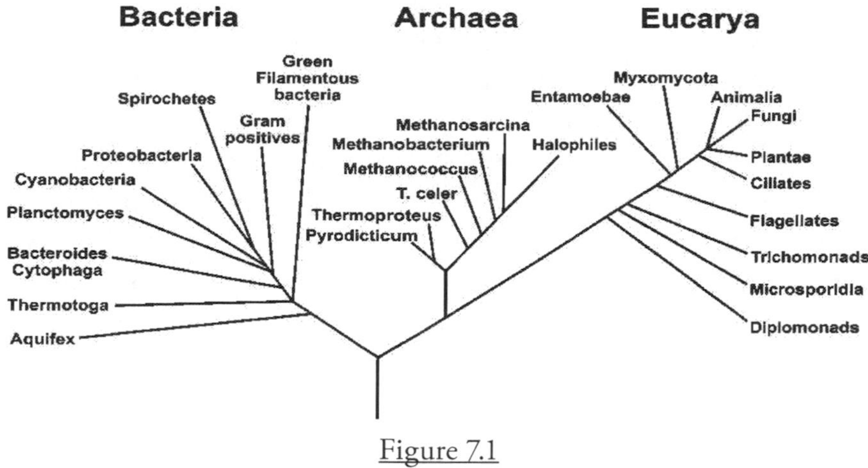

Figure 7.1
The Phylogenetic Tree of Life
Showing the three major divisions

The above developmental steps sequence applies to research at any level in the Catalog of the Universe, from sub-nuclear to the cosmos and including human activities. Although the research progression may seem obvious, I have spelled out the research levels because they define a **scale of maturity** applicable to any science. By this scale, the general pattern is that fields low in the Catalog are the most mature now – having arrived at Level 5, constructing mathematical theories.

Maturity decreases at higher levels, and social sciences may have progressed no higher than Level 1 or 2. This fact should not be surprising or regarded as a fault of the researchers since those working higher in the Catalog have much more complexity to deal with. At higher levels of complexity, repeatable experimental results are harder to obtain. More system variables are present, and environment control, necessary in experiments, may be difficult or impossible. Ethical limitations also come into play at higher levels in the Catalog, setting limits on control and measurements and, therefore, on the reliability of data and theories' significance.

Physics, covering the Catalog's bottom and cosmic levels, has grown up with mathematics from the beginning, and physics education typically includes significant mathematics. On the other hand, chemistry and biology had to wait for mathematics (and computing) to develop higher capabilities and better instruments and experimental techniques to provide more reproducible data. As late as 1990, I heard a prominent researcher in a clinical field say, in a public discussion of a complex body system, that "mathematics does not apply to biology," expressing the feeling that living systems have too much variability to be described by mathematical theories (echoes of "vital force"). Paradoxically, other researchers at the same institution and at the same time were demonstrating reproducibilities of +/-5% or better between experiments on multiple samples of the same tissue type.

Algebra and geometry were invented more than 2000 years ago, and for most of that time, those mathematical systems seemed adequate to describe the natural world. In the next chapter, I will show why in the 16th and 17th centuries, the early scientists realized that their available mathematical methods were incapable of serving as thinking tools for new observations.

CHAPTER 8

The Mathematics of Change

> The only thing that is constant is change. *Heraclitus (540-470 BCE)*

In this chapter, I will explain why ancient Persia and Greece's mathematics were, by the 18th century, inadequate for much of the new scientific thinking. Science needed a new kind of mathematics.

Throughout the reign of gods and the Chain of Being, most humans understood everything in the universe to be perfect, static, and unchanging. The Greek philosopher Heraclitus was an exception, as the leading quote for this chapter shows, but Greek philosophers did not widely accept this view. The assumed result: God created everything that could be made, including every possible being. If anything was missing, the creation would not be perfect. Therefore, all potential species would have to be represented on the earth continuously. Species could not disappear or appear because that would mean the universe was not a perfect creation. In the Chain's human part, God filled every possible society level, from the richest king to the destitute pauper, so there were no empty levels available for those lower in the chain to move up into.

By the 16th century, it had become evident that many forms of change were occurring in the universe. Fossils of creatures no longer living were being uncovered. The list of living species thus had to be recognized as changing over time. Georges-Louis Leclerc, Comte de Buffon (1707-1788), was the first to suggest the evolution of species in his 36 volumes of *Natural History* published from 1749 to 1788 (a century before Darwin published his famous work). There was pressure against the Chain of Being's rigid social hierarchy as people began to strive for upward movement in human culture.

Early forms of mathematics were, not surprisingly, suitable for a static world. The first ideas of algebra were developed in the 9th century BCE by a Persian mathematician, Muhammad ibn Mūsā al-Khwārizmī. Algebra can describe relationships in static or unchanging systems, for example, the forces on a lever. Geometry – the mathematics of shape, size, and space – was developed by Greek philosophers. Geometry examples would be finding areas of shapes such as triangles and circles. Algebra and geometry were adequate to describe static systems and thus for the current concepts of natural phenomena at the time.

Just in the nick of time, a new form of mathematics, designed specifically to describe changing systems, was developed simultaneously in the late 17th century by Isaac Newton and Gottfried Wilhelm Leibnitz. This new mathematics was given the name *calculus*.

The first important idea was to think of rate-of-change – how rapidly or slowly something is changing. As an example of rate-of-change, think of automobile *speed* and its relation to the distance the car covers. We are familiar with the idea that if the speed is higher, the car travels a longer distance in a specified amount of time. The upper graph in Fig. 8.1 shows the car's speed in miles per hour (mph), showing that the driver changes speed three times. We can calculate how much distance the car has covered as time passes (lower graph). 20 mph for 3 minutes (one-twentieth of an hour) means that the car has traveled one mile in that time. In the next 6 minutes, the car covers 6 miles since 60 mph means 1 mile per minute. In the last 6 minutes, the car travels half as much or 3 miles because the speed

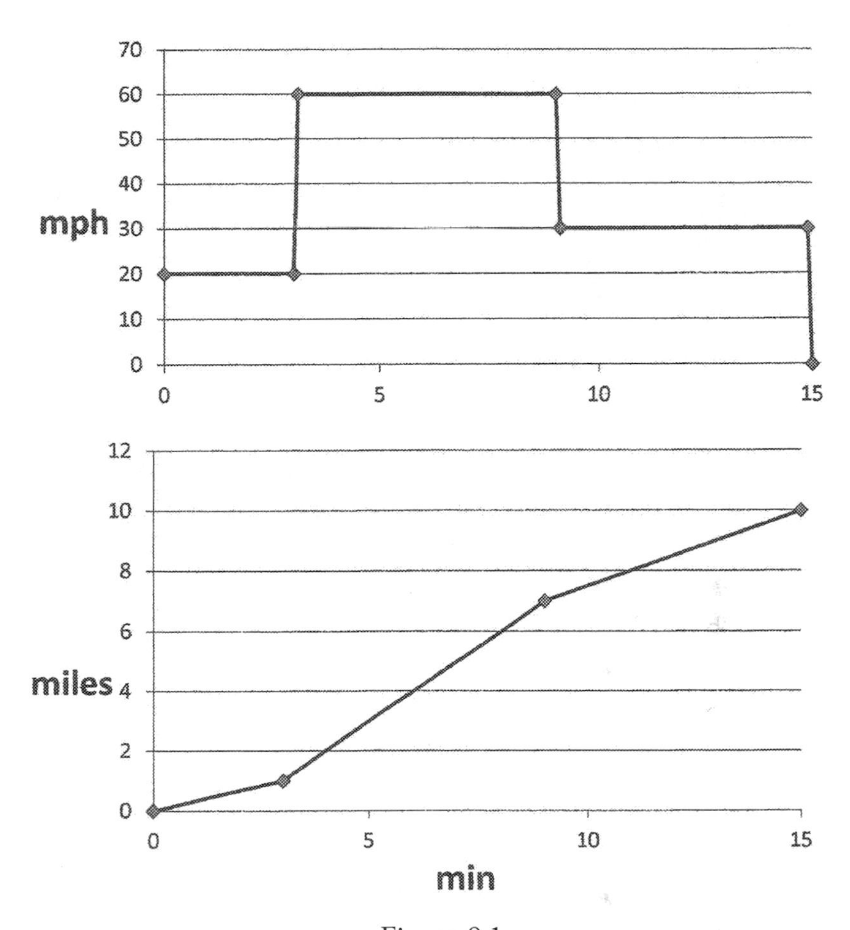

Figure 8.1
The car's speed in miles per hour at the top produces
the distance covered in the bottom graph

was cut in half. With this speed pattern, the car travels a total of 10 miles in 15 minutes. A graphic way of finding the distance covered as time passes is to find the area accumulating under the speed line (upper graph) as time passes. Imagine a vertical line like the vertical axis, identical with that axis at time = 0 min and then moving steadily to the right according to the min scale as if driven by a clock. If we plot a graph of how much area has been covered under the speed line as time passes, we will get the lower graph in Fig. 8.1. Knowing a speed pattern in time, we can calculate the total distance covered by the car.

Imagine that we begin by knowing the distance covered by the car in time (lower graph) and want to find the speed pattern over time (top graph). A straight-line increase in distance means a steady speed. There are three such lines of different steepness, so we need to find each of these regions' speeds. 1 mile in 3 minutes is the same as 20 mph, 6 miles in 6 minutes = 60 mph, and 3 miles in 6 minutes = 30 mph. We have derived the speed (upper) curve from the lower.

The lower graph in Fig 8.1 is the integral or area under the upper graph curve in calculus terminology. The upper graph is the *derivative* or series of slopes of the lower.

The integral calculations for Fig. 8.1 were easy because the car always traveled at a constant speed, so the area under the curve was a simple geometric calculation. Suppose the speed line changed smoothly up and down without sharp changes – a more realistic situation as the driver slows gradually to prepare for curves and speeds up smoothly again after entering

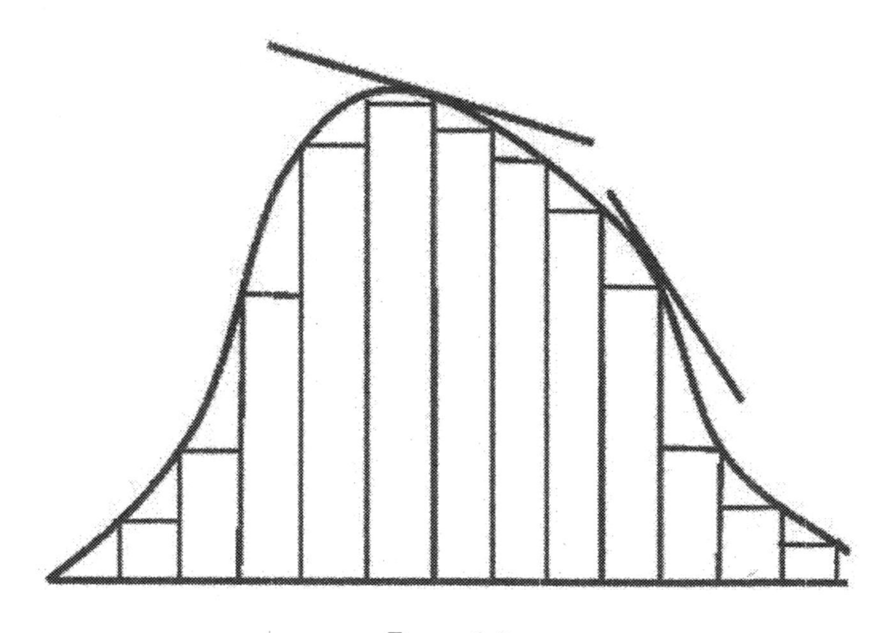

Figure 8.2
An early method of approximating the area under a curve. Also, two tangent lines show the slope of the curve at that point

another straight section? Now how would we get the area under a curving speed line? Newton and Leibnitz's idea was to approximate the area under the speed curve in small steps, as illustrated in Fig. 8.2.[1] Almost two thousand years earlier, Greeks used a similar procedure to calculate the area under a curve by breaking it up into a series of rectangles. Figure 8.2 is a drawing like the one the Greeks used. The total area of all the rectangles is approximately equal to the area under the curve. The approximation gets better as the rectangles' widths are decreased and their number increased to fill the space. Newton and Leibnitz imagined the "limit" as the width of the rectangles was reduced to zero and their number to infinity, showing that at this limit, the multi-rectangle approximation is exactly equal to the area under the curve. Later in the 18th century, Immanuel Kant [2] provided a deeper philosophical understanding of Newton's work in developing the calculus by giving it a metaphysical foundation using a radically transformed version of Leibnitzean metaphysics. [3]

Differentiation (or differential calculus) calculates the rate of change or slope at each instant for a changing system like a moving car. How is this done when the distance record is a curved line? Visually you could imagine laying a ruler on the curve with its edge passing through the point of interest and tipping the ruler to match its slope with that of the curve at the point (see the two examples in Fig. 8.2). For calculation purposes, we can do this at any time, say 5 minutes, by observing the distance covered during a small interval centered on 5 minutes (for example, between 4 1/2 minutes and 5 1/2 minutes). Dividing the distance covered in the time interval by the duration of the interval (1 minute) gives an approximate value for the derivative – the "rate of change of location," otherwise known as "speed," in miles-per-minute, at the 5-minute time. (To convert to the more familiar miles per hour, we would have to multiply the calculated ratio by 60.) If the speed varies continuously, we might have to make our measurement interval smaller instead of being uniform, say 0.1 minutes or even less. Measuring the ratio between distance covered and interval length, we can make the interval smaller and smaller to get an accurate speed estimation even when changing constantly.

Newton and Leibnitz combined these two ideas – differentiation and integration- to be the two fundamental operations of calculus. This new mathematics made it possible to build theories of change.

Newton created a new type of equation as part of his calculus system: differential equations. These equations include variables and *rates of change of variables*, the same thing as variable *derivatives*. If we are interested in the rate of change of a variable y with time t, i.e., the derivative of y with respect to t, we use one of the following symbols

$$\frac{dy}{dt} \text{ or } y'$$

Or if the variable y changed in space (one dimension x) as well as in time t, we could have two *partial derivatives*

$$\frac{\partial y}{\partial x} \text{ and } \frac{\partial y}{\partial t}$$

"partial" signifying derivative in a particular direction

Differential equations are fundamental when it comes to creating scientific models based on relationships. (But don't worry, you will not see equations with these symbols! I will write the relationships out in words.)

Calculus is the basis for some familiar devices. The pump at the gas station is an integrator – integrating (adding up) the liquid flow of gas (the derivative) during the gas flow, giving a total volume of gas sold (the integral). The buyer depends on the device being a good integrator, providing an accurate total volume even when the flow rate changes. Another example is the speedometer that continuously differentiates (finds the rate of change of) the distance covered by the car, thus showing the speed.

With the new mathematics of calculus, investigators could develop theories of changes observed in the natural world. Next, I will describe how to use mathematics to construct a theory.

CHAPTER 9

Developing a Mathematical Theory

> Mathematics is the queen of the sciences. *Carl Friedrich Gauss*

In discussion with philosopher Sam Harris and physicist Max Tegmark, philosopher Rebecca Goldstein described how natural philosophers from the Greeks through the Middle Ages built their theories by intuition – and were wrong. In the 17th century, theory construction began to use mathematics, she said, and modern science's successes began. [1] How, exactly, is a mathematical theory constructed? I will outline the steps in this chapter and describe a simple example.

Expressing a relationship in mathematics means writing one or more equations that describe the interacting factors. The hope is that the equations will reproduce observations with accuracy as high as that of the observations. The equations should also make predictions beyond the observations.

Mathematical theories can be as detailed as necessary to account for the available observations. There is almost always more detail in the real world than needed in a satisfactory theory. Which factors to include is a challenging choice for the researcher. Have too few, and the data are not well accounted for. Include too many, and the solutions can get expensive to calculate and complex to explore. The theory-builder must depend somewhat on intuition to choose what factors to include. It is often beneficial to experiment; start overly simple with few variables and add influences as necessary to account for behaviors seen in the data.

The first step in constructing a mathematical theory for a system is identifying the system's changing properties essential in describing the system's behavior. These changing properties are called *state variables* because they describe the system state through all its changes. Some typical state variables might be the height of an object above ground, the temperature at a point, the concentration of a chemical constituent in a solution, or the electrical potential (voltage) between two points in a circuit. The list of state variables for a system can also include known or hypothetical internal variables, not directly measurable but based on the system structure's clues.

Hopefully, the data the theory is for includes recordings of as many state variables as possible. Ideally, enough data are available so the researcher can verify the theory unambiguously. Too little data, and even though the theory may fit the existing data well, its significance is limited. Mathematical tests are available to ascertain whether measuring only a few state variables is adequate to verify a specific mathematical theory's validity.

A system will also have other properties that remain constant even as the system goes through changes. Examples might be mass or dimensions. These properties, called *parameters*, can also appear in a mathematical theory for the system, but they play different roles from the state variables.

How many state variables are typical for an observed system? The number can vary widely, depending on the nature of the system and the purpose of the study. The performance of a simple model rocket can be described by one state variable: altitude. Adding the rocket's location on the earth's surface at any instant (latitude and longitude) raises the number to three state variables. The attitude and rotation of the rocket itself would add another three state variables (angles). The varying thrust of the rocket engine would add yet another. For a real space launch, the number of state variables will include engine and fuel, and other descriptors and can run into the hundreds or thousands.

The value of each state variable is a number that may vary with time. A record of such a state variable's history can be pictured as a moving point drawing a curve on a graph as time passes. An example would be the temperature at a single point in the air throughout the day. A state variable may also have different values at different locations in space, for example, the force on each point of the surface of an airplane wing. This situation is handled by dividing the wing surface into small areas of defined shape and size, such as tiny rectangles or triangles like early Greeks divided up their curves. A state variable with such a spatial dependence is a *distributed* variable, as it requires many descriptive numbers at each time instant. A distributed variable is equivalent to multiple *scalar* or ordinary state variables. The effective number of state variables can be huge, and even more so when the state variable changes throughout the volume of an object (i.e., in three dimensions), such as the stress in a machine part. In that case, the piece's volume is divided into tiny volume elements (cubes or other three-dimensional shapes), each with a different value for the stress force.

In a changing system, *time* is also an important variable. While other state variables can show changes of any kind, time moves steadily, with or without the experimenter's permission! Time thus plays a unique role in mathematical models and is called the *independent variable* because of its independent behavior. The other state variables are called *dependent variables.*

How does a researcher arrive at the equations for a mathematical theory? The simplest way is to find a known equation that behaves like the observed data to a reasonable accuracy, a curve fitting process. An analogy would be bending a flexible bar to pass through the data points in a graph. Computer programs are available to do curve fitting, offering a choice of equation types and fir accuracy.

Equations from a curve fit are helpful because their values can be calculated rapidly. They are appropriate in training simulations where the purpose is to reproduce known system behavior. However, as mathematical theories for developing system understanding, curve fit equations are not helpful at all. A curve-fit equation is also unlikely to be reliable in predicting new behaviors. The form of the fitted equation most likely has nothing whatsoever to do with the system's internal mechanism. Gaining the scientific benefit of understanding from mathematics requires its use in an entirely different manner.

As emphasized in an earlier chapter, every holon's nature in the Catalog of the Universe results from its internal parts and relationships. To build a mathematical theory, then, it is necessary to start with relationships. Let's take a simple case, a cylindrical bucket that can hold water. The bucket has a small valve at the bottom that can be closed to prevent leakage or opened to let the water flow out. What we can measure – the state variable -- is the *height* of the water in the container. We know the cross-sectional area of the bucket (area of a circle)

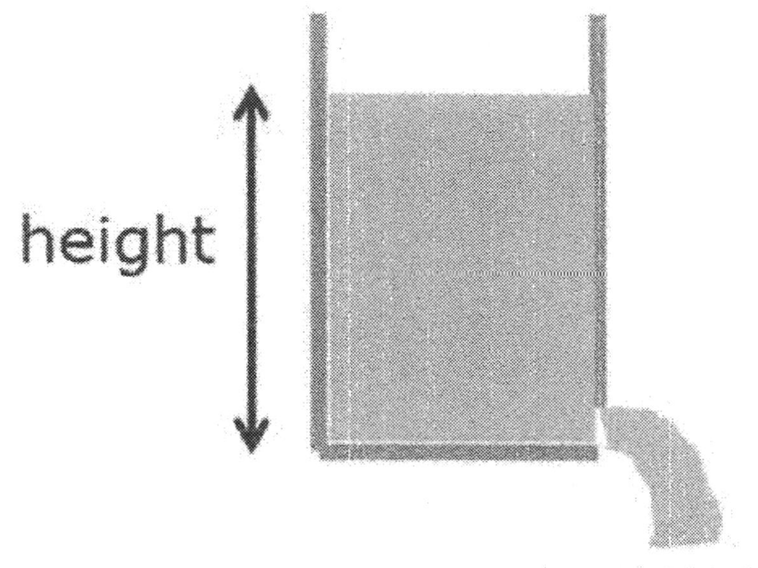

Figure 9.1
Model of a leaky bucket.

and we observe that the instantaneous flow rate of the water out of the valve varies with the height of the water in the bucket: fast outflow when the water level is high and slow flow when the level is low.

The question our mathematical model is supposed to answer is this: Given an initial height of water in the bucket, how long will it take to empty the bucket to a given percentage of the initial height?

What are the important relationships in this example? One governing physical principle is water *conservation*: The water amount that flows out of the valve must exactly match the amount disappearing from the container. (We are not considering evaporation, additional leaks, or someone pouring in more water!) A second principle is that the valve's flow rate is directly proportional to the water pressure at the bottom of the container where the valve is. High pressure at the valve means high flow, and low pressure means low flow. The pressure comes from the weight (pull of gravity) of the water above the valve. It therefore varies with the height of water in the container. Intuition tells us that the water outflow rate will be high initially and decrease as the water level goes down.

To see how these principles can result in a mathematical equation, we must think like Newton and Leibnitz about what happens in a very short interval of time. In a short interval, the volume of water disappearing from the bucket must equal the outflow volume from the valve:

Water lost from the bucket = Outflow from the valve

The water volume disappearing from the bucket is equal to the cross-sectional area of the bucket times a small change in height. Outflow from the valve is equal to the pressure at the valve during the interval times a constant depending on the size of the valve's opening. The pressure at the valve comes from gravity and is equal to a gravity constant times the bucket's water height. The equation that describes what happens in that small interval of time then becomes

(bucket x-section area) * (small change in height) =
- (gravity constant) * (height) * (valve constant) * (small time interval)

An asterisk "*" in the equation means "multiply." The minus sign on the right of the equals sign shows the outflow to be a loss. We can make this equation look less complicated by combining all the constants:

(gravity constant) * (valve constant) / (bucket x-sectional area) = R

Then the equation looks like this:

(small change in height) = - R * (height) * (small change in time)

or in symbolic form

$$\Delta h = - R\, h\, \Delta t$$

where *h* is the *heigh*t of the water in the bucket, *R* is the combined constant, t is time, and Δ in front of a variable means "small change." (Also, in mathematical representation, quantities side-by-side are multiplied together.)

The secret for making this method work is to make the time interval very short – short enough so that we can assume that the change in pressure at the valve is negligible during the interval. We thus have an equation linking two small changes, known as a *difference equation*, or in the limit as the time interval is made smaller and smaller, a *differential equation* for the state variable *height*. Differential equations are relationships between small changes, one of them being a short time interval when the system changes with time.

To summarize, all systems involving change require theories where the relationships are between small changes in different variables. It doesn't matter whether the system is natural or human-made, living or not, or whether the small changes are in time or location or both. The value of developing a mathematical theory for any system is that it requires us to make our theory specific, including the factors that we think are important and omitting those which are not necessary, for example, the bucket material.

Theories based on related small changes are expressed mathematically as differential equations. But we really want to know the system's predicted behavior produced by the specified internal relationships, not just the relationships themselves. Testing a theory described by differential equations – otherwise known as finding the behavior predicted by the equations -- requires that the differential equations be solved, the next chapter's topic.

CHAPTER 10

How Scientific Theories Make Predictions

> God does not care about our mathematical difficulties. He integrates empirically. *Albert Einstein.*

The relation-based differential equation derived for the leaky bucket problem doesn't directly answer our original question about the time to empty the bucket. The equation only calculates the state variable change (water *height* in the container) for a small time step. To answer our question about the bucket system's behavior, we must determine what behaviors are predicted by the differential equation. In mathematical language, the differential equation must be "solved" to answer our question.

Mathematicians have found the complete solution for the leaky bucket differential equation by purely mathematical methods. Thus the full range of behaviors for the water bucket example is well known. Over the last several hundred years, professional mathematicians have found mathematical (or "analytical") solutions to several forms of differential equations, including a few where the state variable depends on one or more space variables as well as time.

Suppose you are a lucky scientist studying a system where the relationships produce one of the solved differential equations. In that case, you can go to a reference book, find the general solution to your differential equation, plug in your parameter values, and you have your complete predicted system behavior! For example, the differential equation derived from our simple water flow problem shows up in many other physical situations. One example is the decay of a radioactive isotope. Analytical solutions are general in that they can produce *all the possible* behavior patterns for a specific differential equation, each behavior resulting from a different set of parameter values.

Now the bad news. Only a very few real-world systems, inorganic or organic and at the lowest levels in the Catalog of the Universe, produce differential equations whose solutions appear in the solved tables. Most systems have qualities that make analytic solutions impossible with any known pure mathematical method. By the 20th century, the inability to find solutions for differential equations of systems under study threatened science's continuing progress. Solving relationship equations is science's way of connecting observations to natural laws. Science's main objective is thwarted without solutions!

There is another way to solve differential equations by manipulating numbers instead of variables. The method works for models with one state variable, as in the leaky bucket example, or many state variables. This method imitates nature directly – letting the relationship equations determine each state variable's fate as time passes in small discrete steps. Although we do not see discrete steps in time or variable values in the real world, Newton and Leibnitz showed that it is appropriate to think of relationships producing changes in variables through large numbers of steps too small to observe. Obtaining accurate solutions with this numerical method may require thousands or even millions of steps. In practice, the scientist chooses a step size by balancing two influences: make the size small enough to obtain the required accuracy vs. make the step size as large as possible to keep the total amount of calculation work no greater than necessary. It is also essential for the scientist to show that the step size itself does not influence the calculated results. The usual method is to cut the step size in half, calculate again, and observe whether the results are different.

Multiple fixed numbers affect the results calculated from the leaky bucket differential equation: the container dimensions, the gravity constant, and valve characteristics when open. These three numbers remain constant throughout any differential equation solution, so for calculating behaviors, we can multiply the three together to create one combined parameter R. The water's initial height in the bucket (the "initial condition") is a second parameter. The researcher must provide numeric values for both parameters to generate one solution.

Two parameters for each state variable are typical for problems of this type. Those problems with spatial dependence (and therefore effectively multiple state variables), like the water in a long and narrow trough, need multiple initial values along a boundary for each spatial direction instead of a single initial value.

To see how the water's height in the bucket (the "behavior" in this example) changes after the valve opens, we can start that height out at a known level and use the difference equation to take many small time steps to see what happens. Mathematicians call this process "solving" the equation, meaning deriving the system's behavior with time from the known relationship. The calculations go like this for the water bucket problem:

1. Set *time* to zero and the water's height to its initial value (Parameter 1).
2. Using the water's height and R (Parameter 3), calculate the water outflow for a small time step.
3. Calculate a new height by subtracting the outflow volume calculated in Step 2 from the volume in the bucket.
4. Advance *time* by the value of the small time step.
5. Repeat Steps 2 through 4, using the result from Step 3 as the new initial condition for *height*, until the desired solution is completed.

The resulting curve (Fig. 10.1) shows how the bucket's water level changes with time for one set of parameter values.

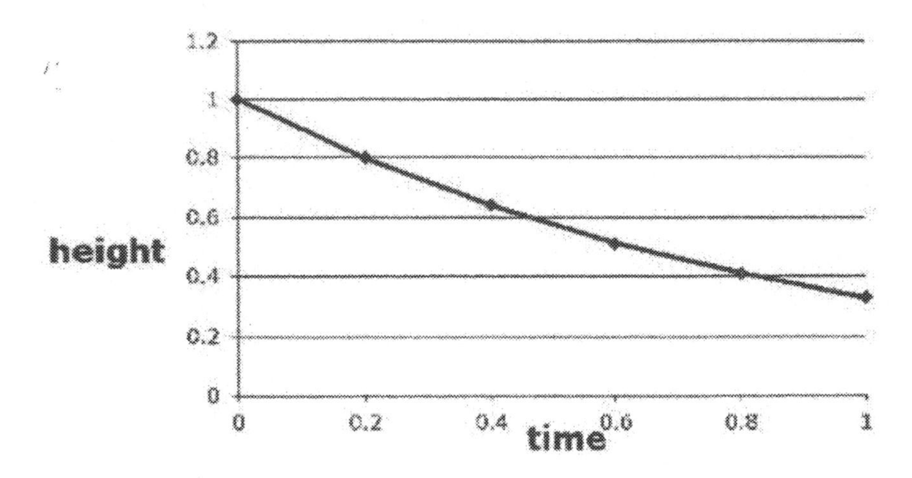

Figure 10.1
Changing water level in the bucket as a function of time after opening the valve.

The water level decreases with time, as we expect. However, it does not go down at a steady rate. Instead, the flow rate gets slower and slower as time passes.

There are two drawbacks to the numerical method of solving differential equations for mathematical theories. The first is that it requires many calculation steps, thousands, or more in a typical problem. The arithmetic in each step must be perfect because an error in one step means that everything from there on is incorrect, i.e., errors are cumulative. The second drawback is that a solution found by the numerical method is valid only for one set of parameter values. For a different initial condition or valve constant, the researcher must repeat the entire step-by-step process. Compare this with the analytic solutions where the researcher simply enters a new initial condition or valve constant into the solution formula.

To review, the mathematical theory process started in Chapter 9 when we used physical relationships to generate a differential equation for the water bucket problem. For a complete analysis of the water bucket problem, a scientist must explore the water bucket's range of behaviors for different combinations of the two parameters' values.

Most real-world systems don't have just one state variable but several or even many. Each state variable will have a separate differential equation.

The researcher must solve the equations for all the problem's state variables simultaneously because it is likely that the state variables will interact. To do that, we leave the relationship equations in difference form and, starting from the initial conditions, use those equations to advance all the state variables together at each time step.

The numerical method for solving differential equations also had applications in the real world. During World War II, artillery teams needed solutions to differential equations to aim their guns. Groups of women using mechanical calculators computed gun-targeting solutions by the tiny step method to be printed in tables for quick lookup by field gun officers. In the 1950s, I learned the technique in a college class. I then had an immediate opportunity to apply that knowledge in a summer job as a student "engineering aide" for a company designing nuclear reactors for power stations. Their design process required the solution of differential equations for reactor design. It was the students' job to calculate those solutions by the numerical method. The other students and I used mechanical desk calculators for the arithmetic and wrote the numbers on pads after each step. Our supervisors would often ask us to repeat an hours-long calculation to make sure there were no errors!

The obvious way to calculate numerical solutions for mathematical theories in the 21st century is to have a computer do repetitive, multi-step calculations. Ready access to significant computing power is a recent thing, however. Electronic computers were developed starting in the 1940s. The first commercial computers were the large multi-cabinet, room-filling behemoths designed for business data processing. Operators entered calculation instructions in a stack of punched cards. Results appeared as numbers on accordion stacks of green-striped paper. Later, humans communicated with computers through remote terminals. Programmers wrote instructions in FORTRAN ("formula translator"), a programming language designed to describe calculations. These large computers were so expensive that only major companies, national laboratories, and a few universities could afford them.

As a graduate student in physics in the early 1960s, I was fortunate to have access via punched cards to an IBM 7070 computer on the Brown University campus. The 7070 was a room-sized machine with computing power far less than the processor in a cheap cell phone in 2021. Using this resource, I developed and exercised a mathematical model for a nuclear particle counter, answering some counter functionality questions. This work earned me a master's degree in 1964.

The first computer small enough in size and price to be in a research laboratory was the PDP-8 produced by Digital Equipment Corporation (DEC). The original model sold for $18,500 in 1965 and was about the size of a small refrigerator. The PDP-8 went through several stages of development. A version specially adapted for experiment control and data recording was labeled PDP-12. DEC replaced the PDP-8 and 12 with the PDP-11 and VAX with larger word sizes and faster compute speeds. Compaq acquired the company in 1998.

The introduction of the IBM Personal Computer in 1981 marked the beginning of computing that an individual researcher or a small office could afford. With the need for extra space and staff support gone, a scientist or student could have full-time interactive access to significant computing power. However, scientific computing did not blossom immediately as two more developments were needed: new software to make mathematical theory more accessible and an appreciation for the contribution of computed mathematical theory to science.

Early in the 21st century, computers and software have become essential tools for science, both for observations and theories. Computers control experiments and record, organize, and display the data. Even more important, computers now implement scientific theories. Computer solutions now lift the limit to the complexity of scientific theories resulting from analytic methods' limited ability to solve differential equations. More accurately, the size and speed of available computers now limit scientific theories' complexity that researchers can investigate. Instead of just a few state variables or an inaccurate version of the relationships between them, some current theories contain a million or more state variables. The practical limit is likely to keep rising. The price paid for this expanded theoretical capability is that each computer solution to a set of theory equations is specific to one unique set of conditions, saying nothing about the system's behavior under different conditions. This characteristic of theory by computer has opened a new branch of science that I will discuss later.

Theories built by reduction (explanations coming from lower Catalog levels) use computer solutions to verify that the hypothetical system components, when linked together by assumed relationships, show behaviors comparable to those recorded from the original system. This method of understanding by reduction is now standard not only in physics but also in chemistry/biochemistry, cell physiology, organ physiology, organ system physiology, and even living system biology and ecology examples. Yes, mathematics does apply to biology, and it significantly aids understanding! As we will see in a later chapter, there are still levels of the Catalog of the Universe where mathematics currently plays little or no role.

There is another method of using computers to aid science that does not involve differential equations and their solutions. That method is Artificial Intelligence or AI, a sophisticated type of automatic pattern recognition. AI is rapidly becoming a standard tool for businesses with large customer lists. AI can help a company find patterns in large groups of customers' behavior, permitting marketing optimization. AI can discover patterns in an individual customer's purchase history so that marketing can be tailored to each customer's interests. For example, a music streaming service uses AI to learn each customer's music preferences in great detail, right down to the type of instruments the customer prefers to listen to. The company then can suggest new selections or albums out of the millions available to match the customer's preferences.

Artificial Intelligence is created by programming a digital computer to imitate a linked network of biological nerve cells, creating an Artificial Neural Network (ANN). The ANN is "trained" by presenting it with a large number – tens of thousands up to millions or more – of identified examples of what is to be recognized, automatically adjusting the internal connectivity in the neural network until the accuracy of recognition reaches a satisfactory level. ANNs automate facial and voice recognition, for example, and text analysis. Artificial Intelligence can be helpful in science by identifying patterns in any data, including pictures. For example, smartphone apps can identify plants and flowers viewed by the phone's camera.

The method of organizing scientific observations and describing them in mathematical models presented in the preceding chapters is also pattern recognition. However, there is a fundamental difference between the significance of results from the mathematical and ANN methods. The mathematical modeling method offers the possibility of a mechanistic explanation of the observations, an explanation in terms of more fundamental processes or laws. While AI can be very discriminating in its pattern recognition within a large dataset, the method offers no understanding of the processes producing the recorded data. Developers of ANN have admitted that they do not know how ANN's work. They have no theory that would enable them to take the detailed structure of a trained ANN and derive what the ANN is trained to recognize. [1]

In summary, scientific research is now done by teams using sophisticated instruments and computers to record large quantities of data from a complex system. Understanding all this data is then aided by constructing mathematical theories and implementing them in computers. The level of detail possible is determined by the amount and quality of available data, the complexity of the mathematics in the theories, and the size and speed of computers available to solve the theories' equations.

In the next chapter, I will describe how computers began to contribute to theory development in science.

CHAPTER 11

Scientific Theories Need Computers

> Science is what we understand well enough to explain to a computer. Art is everything else we do. *Donald Knuth.*

As I elaborated in the previous chapter, computers provide a way past the barrier of solving differential equations, with suitable hardware and software becoming available just as needed to keep science advancing. Two landmark research publications, one in biology and one in mathematics, marked the beginning of computer calculation contributions to scientific theory. The first example shows how mathematics and computing made it possible to decipher a complex biological process mechanism: the electrical action potential generated by a nerve as an essential part of its information processing or transmission function.

The knowledge that electricity was involved in biology began at the end of the 18th century when Luigi Galvani observed that when an electrically charged scalpel touched a dead frog's leg, the leg muscle contracted. By the middle of the next century, researchers learned that nerves generate electrical activity in the form of a propagated impulse named an "action potential."

In the late 1940s, three researchers at the University of Cambridge, A. L. Hodgkin, A. F. Huxley, and B. Katz were determined to decipher the action potential mechanism in nerves and how it propagates. The experimental apparatus they needed had just been designed in 1947 by an American scientist Kenneth S. "Kacey" Cole. He began with a "large" nerve cell axon (0.5 mm diameter, like fine pencil lead) from a giant squid, and he inserted two fine silver wires down the inside of this axon. The saltwater bath on the outside was the opposing electrode. Cole could control or "clamp" the electrical potential across the axon membrane with an electronic feedback circuit and measure the ionic current through the membrane. [1]

Hodgkin, Huxley, and Katz used Cole's apparatus design to measure the individual ion currents through the axon membrane and how those currents varied with the electrical potential across the membrane. [2] They found two currents to be important: one of sodium ions and one of potassium ions, along with a small residual current of miscellaneous ions. Hodgkin and Huxley generated mathematical equations that described the relationship between current and voltage for both ions and the residual from the experimental results. [3] Figure 11.1 is a summary of their results, shown as an electrical circuit.

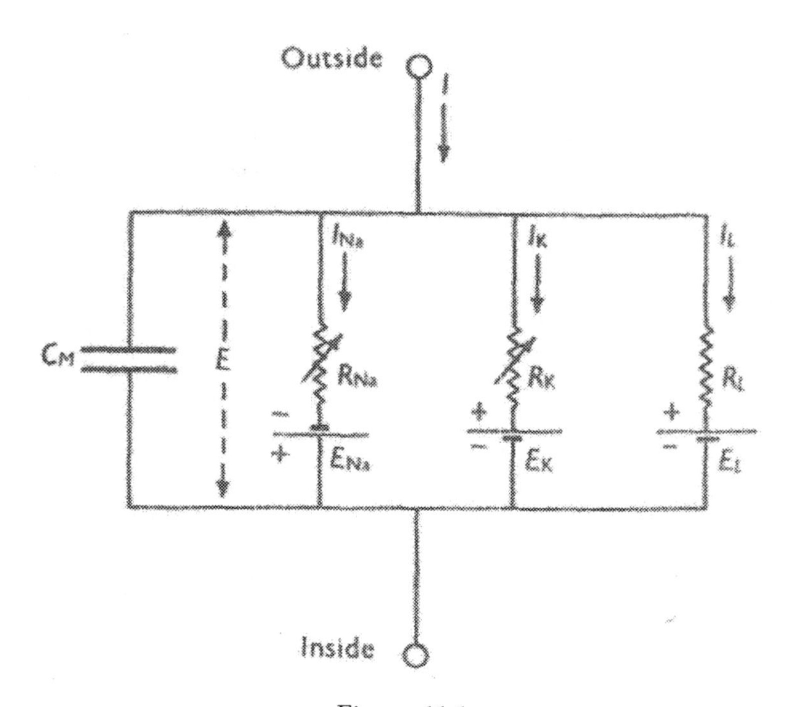

Figure 11.1

An electrical representation of Hodgkin and Huxley's three-ion current model for the squid axon as flows of the three ion types, [3]

The three R components represent the cell membrane's variable conductances to the sodium and potassium ions and the mixed remainder.

The E components represent the effective "batteries" generated by the different ion concentrations inside and outside the cell membrane. For example, the potassium concentration outside the axon is ordinarily low, and it is much higher inside the axon, so the negative terminal of this "battery" is on the inside. The reverse is true for the sodium "battery." The fixed C component represents the capacitance produced by conducting ion solutions on either side of the insulating cell membrane. The equation describing the current through a capacitor is a differential equation, while the equations for the ionic currents are complex but not differential. Three ion currents and a capacitive current share the same electrical potential between the inside and outside of the cell. Hodgkin and Huxley had performed a reduction analysis of an essential biological system.

Now it was time for the critical test: Would those separately measured components, when linked together, reproduce the observed behavior of the action potential in the nerve axon? Hodgkin and Huxley had equations for four state variables to describe the four branches of the circuit. The nature of the differential equations for their system made solution impossible by analytic methods. There were no laboratory or desktop computers in the 1940s, so the researchers applied for some time on an early experimental computer. Before their scheduled time came up, engineers took the experimental computer down for three months for additional development.

Hodgkin and Huxley decided to do the computation manually using a hand-cranked mechanical calculator for the arithmetic rather than waiting for the experimental computer. They battled some instabilities in their calculations, and by the time they achieved the results they hoped for, a month had gone by. But, their work was a success! The solution to their differential equations reproduced the nerve's electrical impulse very accurately! (See Figure 11.2) They demonstrated that the three ion current systems working together did indeed account for the electrical impulse's behavior.

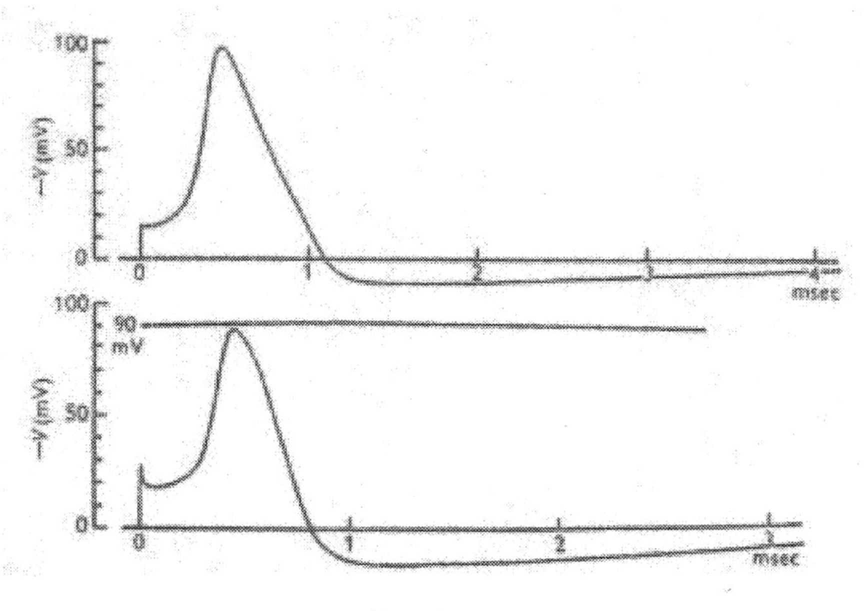

<u>Figure 11.2</u>
Lower: Squid nerve action potential recorded by Hodgkin and Huxley. Upper: Action potential generated by solving the ion current equations developed by Hodgkin and Huxley. [3]

In the language of the Catalog of the Universe, Hodgkin and Huxley's accomplishment was to build a successful theory for a nerve. They first reduced their system into components (looking one level down the Catalog for understanding) and how they were linked. Then they verified the correctness of their theory by reconstructing their system with a mathematical theory solved by human computation. Their paper is a model of how researchers should work – reduction followed by synthesis. This work has inspired many students and similar investigations for other electrically active tissues, such as skeletal and heart muscle. The experiments and papers also demonstrated that a complex biological system could be analyzed successfully by physical methods without reference to "vitalism."

In this example, the human researchers calculated the solution for their model equations by the same steps that a computer would have carried out had one been available. It required a month of human effort to simulate one-hundredth of a second of actual nerve time! Today, a modest laptop computer can calculate the same solution in about the same time the nerve takes to accomplish it – a few thousandths of a second.

The second example of the new computer approach was published in a Los Alamos National Laboratory report in 1955 by Enrico Fermi, John Pasta, and Stanislaw Ulam. Some years before, Fermi had posed a problem in mathematics, looking for a solution to a particular type of differential equations. In their report, the authors showed how Fermi's equations could be solved numerically by a computer, and they explored the range of possible solutions. [4] This report has been lauded as the earliest example of a computer finding numerical solutions to investigate a problem in mathematics. A 2009 review in *American Scientist* said of this report: "It is not an exaggeration to say that the FPU problem, as the system Fermi, Pasta and Ulam studied and is now universally called, sparked a revolution in modern science."

Computer solutions of mathematical theories now play a significant role in science. However, the nature of these solutions has caused a considerable shift in scientific theory. The next chapter will show why building scientific theories with computers has caused a significant change in theory testing.

CHAPTER 12

Theories Generate Many Possibilities

> Life is a paradise of endless possibilities! *Mehmet Murat ildan*

While computers can find solutions to theory equations that mathematical methods cannot solve, the computer method introduces a new challenge.

To understand this challenge, think about the leaky bucket example in an earlier chapter. This example had one state variable (the varying height of the water) and two parameters, the initial height of the water and the combined constant with the bucket, valve, and gravity information. In theory, at least, every different combination of these parameters' values will produce another possible behavior of the system. How should the range of values for both parameters be tested to investigate the full range of system behaviors? One way would be to choose a fixed number of values for each parameter, say 20, and try all (20 x 20) 400 possible combinations. In the case of a theory with two parameters, that would work quite well. But this approach does not work for more extensive models with multiple variables (and thus more parameters) because the number of combinations quickly gets too large. Because the changes produced by varying one parameter depend on all the other parameters' values, any parameter can (at least in theory) affect any variable.

Trying various combinations of parameters to study the range of behaviors of a theory is called "exploring parameter space." For the leaky bucket example, the parameter space is a two-dimensional surface with values for one parameter measured along the horizontal axis and the other along the vertical axis. Any point on this surface represents a pair of numbers, one value for each theory parameter. "Exploration of the parameter space" means calculating model solutions at multiple points on the surface to show the theory's range of behaviors.

For example, we could make a grid of values, say 11 values for each parameter, for a total of 11x11=121 points, each point representing a pair of parameter values (Figure 12.1). The computer then calculates a solution to the leaky bucket problem at each grid point.

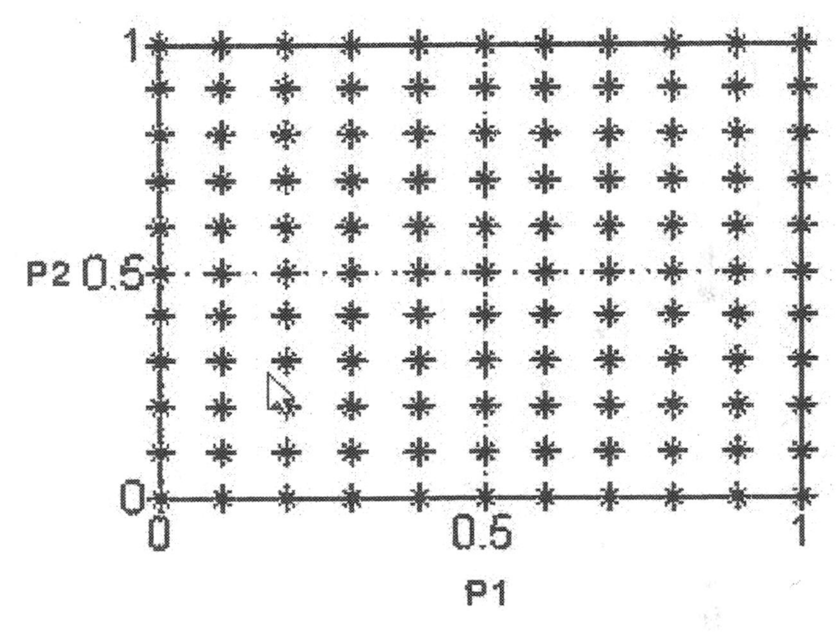

Figure 12.1
A two-parameter space covered by a grid of points,
Each representing a pair of parameter values (P1, P2)

Realistic scientific models usually have more than two parameters, meaning that parameter space is no longer a two-dimensional surface but a "volume" in a high-dimensional "space." A space with more than three dimensions is hard to visualize, but computers can keep track of large sets of parameters, even if visualization is impossible. Exploring a high dimension space by testing all possible combinations becomes impossible, though, as the following example illustrates.

Consider an example from my research into mechanisms of the electrical activity of heart cells. Like all other cell types, each heart cell is enclosed in an extremely thin electrically insulating membrane made up of two layers of lipid (fatty acid) molecules back to back, called a lipid bilayer. The environments inside and outside the cell consist of water that floats many kinds of molecules and atoms. Some floating particles are electrically charged, having one or more electrons removed or added, leaving them as ions with positive or negative charges. For electrically active cells such as heart muscle, skeletal muscle, and nerve cells, the three most essential ions are sodium and potassium, each with one electron missing (symbolized by Na^+ and K^+) and calcium with two electrons missing (Ca^{++}). Concentrations of Na^+ and Ca^{++} are relatively high outside the cells and low inside. The concentration of K^+ is relatively low on the outside and high on the inside.

Some uncharged molecules can float right through the lipid bilayer membrane in either direction, but others and all charged molecules and ions cannot. There are unique transporter molecules embedded in the cell membrane and facing the solutions on both sides. Each transporter can move one or more types of ions or other molecules across the membrane when the conditions are right. Researchers have identified more than a dozen of these specialized transporter molecules in heart cells, for example.

My colleagues and I studied several of these transporter systems, including a sodium-calcium (Na^+-Ca^{++}) exchanger. This transporter is like a revolving door, with one Ca^{++} ion moving in one direction across the cell membrane and three Na^+ ions in the other direction. This "door" does not turn, moving the ions and letting them get off on the other side unless all four ions are in place. The direction the "door" revolves depends on the ion concentrations on each side of the membrane and the electrical potential across the membrane.

To build a mathematical model for the Na^+-Ca^{++} exchanger, we thought of the exchanger as a multi-step chemical reaction. Figure 14.2 shows one of the possible reaction schemes we considered (neither the simplest nor the most complex). [1] E and E' represent the exchanger molecule facing the two directions. To understand how the transporter works, start with the E at the top of the diagram. Follow the steps down as the four ions bind on in any order to reach the middle of the diagram. Then the loaded exchanger flips from E to E'. Continuing down the diagram, the four ions get off in any order leaving E'. Finally, the E' form cycles back around to the empty E at the top. A single exchanger molecule can repeat this cycle up to 100,000 times per second.

This diagram has 16 states (their concentrations being state variables) and 22 reaction steps coupling the states. Writing the differential equations and the computer code for these relationships is tedious but straightforward. Each reaction step requires two rate coefficients or parameters (one forward and one backward rate coefficient) with values for a calculation. Other essential numbers for each calculation include ion concentrations inside and outside (4 numbers) and the membrane's electrical potential. Together, 44+4+1=49 numbers (parameter values) are required to calculate what happens in this chemical reaction. One of these values, usually an outside concentration or the electrical potential, is varied in an experiment. That still leaves 48 parameters to be tested -- a very high dimensional space!

How can the behavior of this theoretical reaction be investigated? The researcher could use physiological values from measurements for the two inside and outside ion concentrations and electrical potential, but that still leaves 44 rate coefficients. Mathematically, it is not adequate to vary one rate coefficient at a time, leaving the other 43 fixed because the reaction steps are not independent but change together like the many springs in a depressed mattress. What about testing combinations of the 44 rate coefficients, say ten values for each one? There would be ten to the 44th power combinations to test. If our fast computer could check a million combinations a second, our investigation would take 10^{38} seconds or three times 10^{30} years – vastly longer than the longest estimates of the age of the universe! We clearly could not even *think* of investigating all possible parameter combinations! The problem arises because testing all possible combinations requires multiplying together the number of possibilities for each parameter.

Scientists studying complex systems have some methods to find solutions to problems with high-dimension parameter spaces. One technique, called "pruning," is to eliminate regions of the parameter space where some physical law like conservation of energy would be broken. Another method is to start with one known solution and investigate the "parameter space" close to that solution. Using such methods, we found an approximate way to explore a reasonable range of behaviors for our sodium-calcium exchange model, finally restricting our search to a 3-dimensional parameter space that we could search more completely.

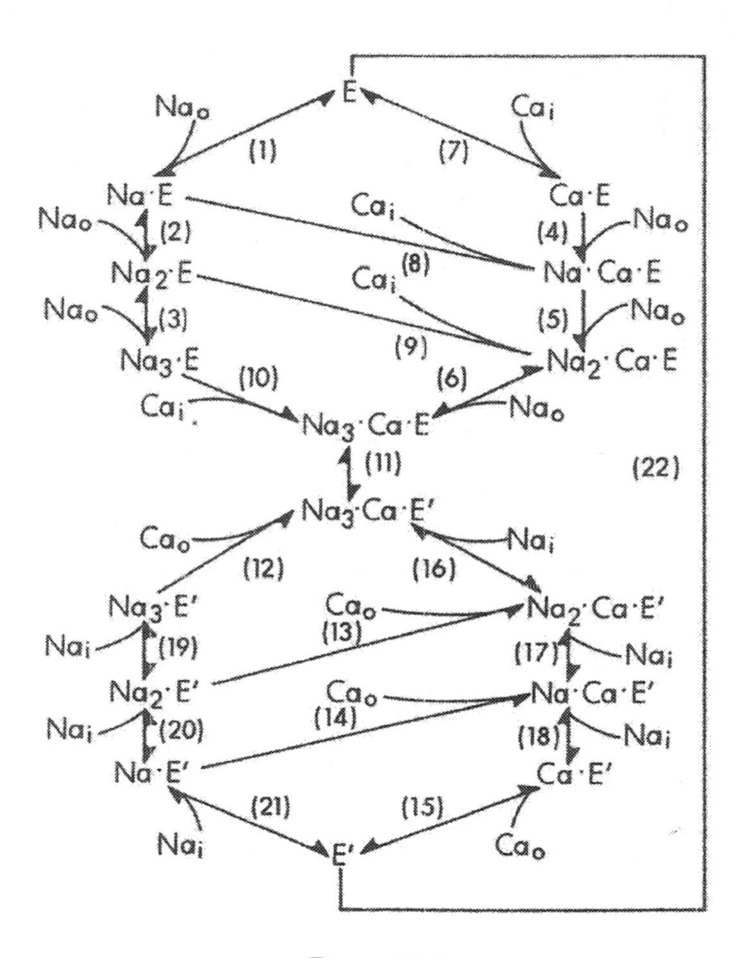

Figure 12.2

Chemical reaction diagram for a theoretical model of a sodium-calcium exchanger. [1]

In a later and more complex model for this same transporter, we did explore a 16-dimension parameter space. How do you study a space with many dimensions? The answer is to find or calculate a "value" number for each solution that describes how "good" the solution is. An example would be the error between the solution and some experimental data. Then, the computer can search through the multi-dimensional space to find the lowest value for that error. As a visual example, consider a simple model with two parameters. The grid at the base of Figure 12.3 represents pairs of values for the two parameters (like Figure 12.1). Error results appear as heights above the base grid points. The result is a surface with many kinds of features: peaks, valleys, slopes, flat areas, cliffs, etc.

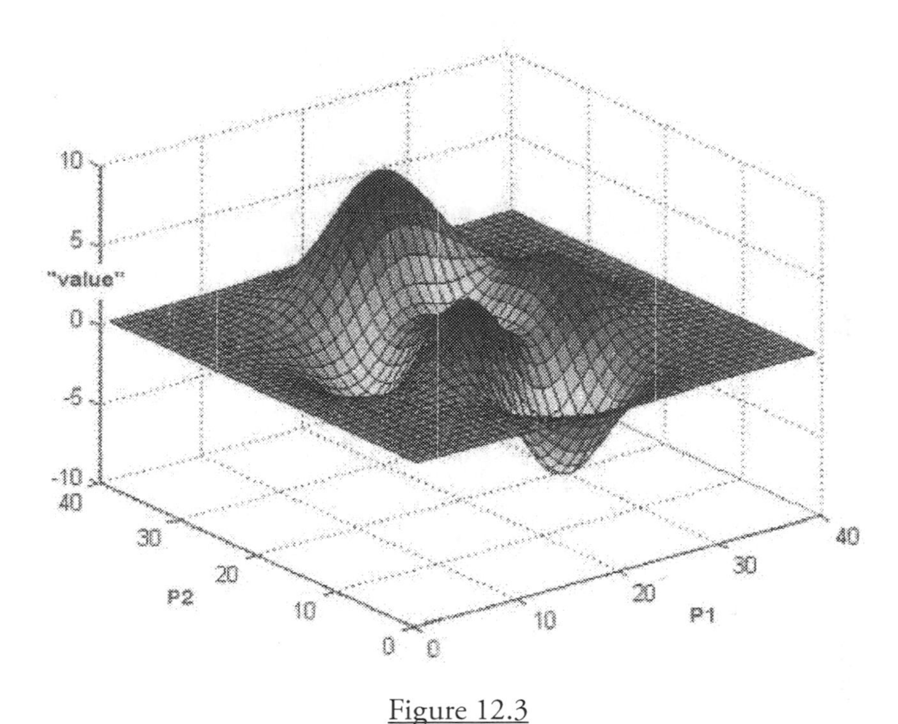

Figure 12.3

"value" surface for a model with two parameters P1 and P2.

There is a specialized area of research where the aim is to find fast and automatic methods for finding a given type of feature, say the highest peak to optimize an industrial process, for example, or the bottom of a pit giving the lowest error. [2,3] These search methods work in many-dimensional spaces not readily visualized. The automatic methods are not foolproof because they cannot explore every region of the parameter space. Sometimes a search can get fooled by a "local best" that is not the "overall best." The automatic methods are still limited for vast numbers of parameters, say above 25, because too many model equations are required.

Therefore, it is possible to develop a mathematical theory, i.e., write the relationship equations, for a system with many state variables and parameters, and not know all the theory predicts!

One way to get around this problem is to measure some or all the actual system parameters in the natural world. For example, in the earlier leaky bucket example, it might be possible to obtain the valve constant from the manufacturer or measure it directly. Then the problem of exploring a large parameter space can be reduced or even eliminated. Such measurements may not always be possible, however.

We thus were able to produce a scientifically credible mathematical model for this vital ion transport molecule by combining some experimental data from biochemistry and ideas from chemical reaction theory. Exploring this model required a computer to solve the model equations, some experimental data, theoretical concepts from a lower level in the Catalog, and special computer programs to help us explore the model's behavior.

There is another challenge that can occur with some mathematical models of complex systems. Investigators expect that making a small change to model input causes a correspondingly small change in the output. For some systems, this is not the case; a tiny disturbance at the input produces a totally different output. Edward Lorenz first noticed this unexpected behavior in 1961 as he was attempting to build a computer model to predict the weather. [4] His computer was not capable of simulating worldwide weather. Still, Lorenz's model did show some realistic behavior in a small square area. There were no clouds in his square and it never rained, but the winds did change direction and speed and an occasional cyclone would appear. Every minute the computer printer would spit out a new set of numbers describing the new pattern for the next day. (There were no graphic computer displays in 1961!)

One day, Lorenz decided to explore the behavior in a part of one calculation in more detail. Instead of starting over from the beginning, he typed in as initial conditions the calculated numbers from the middle of the previous calculation run. He expected that the results would repeat the last part of the previous calculations, but they did not! After a short period in which the new results almost matched the original, suddenly the output changed to be completely different! Lorenz, of course, checked the computer to make sure it was working. He finally realized that when he entered the intermediate results from the first prediction, he entered the value for each point on his weather map with fewer digits than the computer was saving internally when it did the original forecast. Thus, the rerun's input conditions were *almost* the same as the values from the middle of the earlier run, but not *exactly* the same. His model was very sensitive to the input valucs.

Lorenz's discovery has been called the "butterfly effect." The flap of a butterfly's wings days or weeks ago might affect today's weather. Mathematicians call this effect "chaos," and it is not limited to weather models. Models whose equations contain time delays or nonlinearities (model variables multiplied or divided rather than added or subtracted) are likely to show chaos under some conditions.

An example of a system with delays is hematopoiesis – the regulation of white and red blood cells and platelets in the human body. [5] These blood elements enter the bloodstream from bone marrow. When the blood's oxygen level decreases, a substance is released that causes more blood elements to be released from the marrow. There is a production delay between when additional components are needed and when they arrive in the circulation. Under normal conditions, the cell density of these elements in the blood oscillates slowly around a normal value with a period of about 20 days. There is a disease condition where this control system becomes chaotic, and the white blood cell count oscillates widely with a period of about 72 days.

Lorenz discovered one simple nonlinear population model that displays chaos. That model now has Lorenz's name. [6] It is written as a nonlinear equation in one variable u:

$$u_{t+1} = u_t \exp[r(1 - u_t)]$$

Figure 12.4 shows some results from the Lorentz model. When the growth parameter r is less than 2, the output stays constant (not shown in the figure). When $r = 2$, the output oscillates with a steady amplitude and frequency (panel a). For r values of 2.6 (panel b), 2.671 (panel c), and 3.116 (panel e), the output still oscillates, but the pattern is different for each r value. For r values of 2.9 and 3.5, the output oscillates, but the pattern is irregular – chaos.

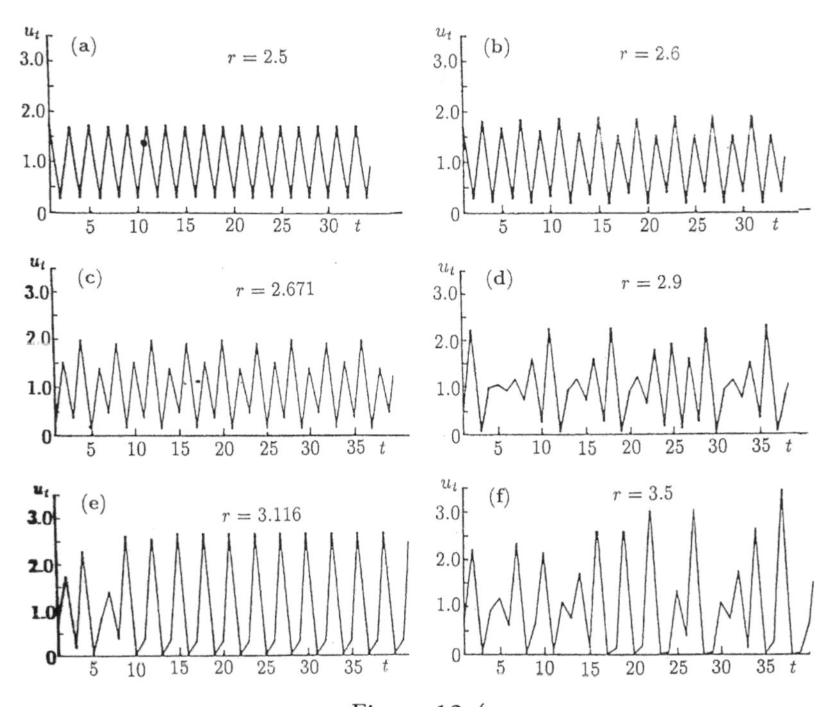

Figure 12.4

Results from the Lorenz population model for different values of the growth parameter *r*. [7]

In the years following the publication of Lorenz's paper (1963), his equation found application in many fields of natural science. For example, Robert M. May wrote a detailed review article for *Nature* in 1976 describing the equation and its behavior for ecologists' population studies.[8] The equation was included in the popular text *Ecology* by Robert E. Ricklefs and Gary L. Miller in their chapter on "Population Fluctuations and Cycles" under the name "Ricker Model." [9]

Remember that this "chaotic" behavior is not caused by lousy computer programming or an error in the computer hardware. Nor is there any apparent random process in the model; the model is entirely deterministic. Yet, it displays some output that has no evident pattern.

Because of the parameter space challenge described in this chapter, working with mathematical theories implemented in computers has created a new scientific research mode. That new research mode is described in the next chapter.

CHAPTER 13

The Third Branch of Science

> Exploration is wired into our brains. If we can see the horizon, we want to know what's beyond. *Buzz Aldrin.*

Traditionally, activities in any scientific field have been divided into two camps -- experiment and theory – frequently with different people doing the two kinds of work. Mathematics is the widely accepted theoretical platform for science, but by the middle of the 20th century, some theoreticians realized that they had a problem. Analytic mathematics was an inadequate tool for studying anything beyond the simplest systems. It was not too difficult to construct the differential equations for a complex system based on relationships. However, the theoretical process and science progress were incomplete until researchers could find solutions to their relationship equations.

It is a lot easier to generate a theory in any study area than to figure out what the theory predicts. When scientific theories created only differential equations that analytic methods could solve, scientists could look at the solutions to those equations (the solutions were also equations) and see the entire range of behaviors predicted by that theory. For example, they could vary one parameter at a time to see what effect that parameter had on the solutions. The drawback was that researchers expressed very few complex system theories in mathematical form because solutions were available for so few differential equation systems representing real-world models. Intuition thus played a significant role in comparing theories. Intuition is not only unreliable but also variable between individuals. Fields of high complexity like biology and sociology didn't even attempt mathematical ideas because their relationship equations were too complex to solve by analytic methods.

As we saw in the previous chapter, when a scientific theory is implemented numerically in a computer, the idea must be probed and tested. The procedure tests different conditions (sets of parameter values) to see what behaviors the theory predicts and whether they match the experimental observations. Thus was born the Third Branch of Science – Experimental Mathematics or Experimental Theory, a combination of theory and experiment. The researcher explores theoretical equations by solving them with different sets of parameter values. This form of exploration by computer is essential for comparing theoretical results with experimental observations. The process is also valuable for planning additional experiments and measurements so the correct data can be acquired to best distinguish between competing theories.

Other writers have also commented on the significance of computer-implemented theories for complex systems. Physicist and philosopher of science Heinz Pagels wrote a book in 1988 on this subject entitled *Dreams of Reason: The Computer and the Rise of the Sciences of Complexity.* [1] Pagels described computer modeling as a new way to do "experiments" and explained a computational view of physical processes:

> "According to [the] computational viewpoint, the laws of nature are algorithms that control the development of the system in time, just as real programs do for computers. For example, the planets, in moving around the sun, are doing analogue computations of the laws of Newton." [2]

Pagels also quotes Peter Lax, a mathematician at the Courant Institute of New York University, describing computer modeling as a new branch of science:

> "The traditional branches of science, the experimental and the theoretical, correspond to the traditional sources of knowledge. In the last two decades, a third branch, the computational, has joined the other two and is rapidly approaching its older sisters in importance and intellectual respectability." [3]

Pagels himself went on to make a significant prediction:

> "I am convinced that the societies that master the new sciences of complexity and can convert that knowledge into new products and forms of social organization will become the cultural, economic, and military superpowers of the next century." [4]

Four years after Pagels' book was published, the April 3, 1992, edition of *Science* included a news perspective entitled "The Third Branch of Science Debuts." The article had the following subtitle: "Computer simulation has opened a new eye on the world, giving scientists in fields from biology to high-energy physics a way to perform experiments that would otherwise be impossible." [5]

Vinton Cerf, widely recognized as the Father of the Internet, wrote:

> "We are reaching an exciting period in scientific discovery in which computation is as important as laboratory experiment and observation. We can invent our own universes and test them for compatibility with the real one we can measure. Indeed, we may find that our predictions could draw our attention to phenomena we might never have looked for, were it not for the revelation of computation." [6]

In 1997 John Horgan, a senior writer at *Scientific American*, published a book with the title *The End of Science: Facing the Limits of Knowledge in the Twilight of the Scientific Age.* [7] Horgan was stimulated to write his book by a conversation with Roger Penrose, a British physicist. A month after talking with Penrose, Horgan attended a symposium with the title "The End of Science." Penrose was releasing his book *The Emperor's New Mind* [8], in which he considered the scientific puzzle of consciousness. He had searched all science fields for methods of understanding consciousness and concluded that science could not possibly account for this mysterious mental phenomenon.

As it turned out, the symposium addressed the premise that *belief* in science, rather than science itself, was coming to an end. However, at the meeting, Horgan listened to a presentation by Gunther Stent, a biologist from the University of California, Berkeley, who had for years promoted the idea that science itself was facing an end – because it worked so well. Twenty years earlier, Stent had published *The Coming of the Golden Age: A View of the End of Progress* [9], in which he predicted that not only science but also technology development and the arts would be ending. Stent based his prediction on a law of acceleration proposed by historian Henry Adams at the turn of the 20th century. Adams made his proposal because he understood that not only was science advancing, but its rate of advance was also increasing because new science generated new scientific tools. Stent understood that any process with exponential growth cannot continue to expand indefinitely and thus must end.

As we have seen, science that depended on analytic mathematics theories did virtually end by the middle of the 20th century. Fortunately, computers came along just in time to offer a new type of theory, not one that provides the complete spectrum of predicted behaviors of the investigated system but one that mechanizes all the system components and relationships. Science thus did not end but changed its cycle of activities.

→ Observe (real) → Theorize → Experiment (real) ↩

has been replaced by

→ Observe (real) → Theorize → Experiment (theory) → Experiment (real) ↩

The problematic step of solving differential equations has been mechanized and emphasized instead of identifying the dominant mechanisms.

Combining elements into a system and exploring that system's behavior through computer calculations has been essential in validating theories constructed by reduction. That is where a system has been taken apart experimentally and reassembled for testing. This same capability now makes it possible to create something new by linking holons from a given level to see what might result, i.e., moving *up* the Catalog. The next chapter is about increasing complexity by moving up the Catalog hierarchy, a process opposite to reduction and called "synthesis," producing "emergent" results.

CHAPTER 14

Emergence Is Creative

> Life is not what you expect: it is made up of the most unexpected twists and turns. *Ilaiyaraaja.*

The traditional scientific method has been *reduction: taking a system apart to understand* how the system works. The opposite approach is to put things together: *synthesis*. Synthesis is helpful as the final step after a reduction to reassemble the identified components or sub-holons and their relationships, whether by experiment or mathematics. But the importance of synthesis goes far beyond validating reduction.

Experimental synthesis is possible in chemistry, material science, and a few other areas, but attempts have only begun in biology. Scientists created the first artificial "cells" in the 1960s. These were small objects about the size of biological cells surrounded by permeable, ultrathin membranes of nylon, collodion, or cross-linked proteins. The artificial cells were filled with enzymes, hemoglobin, proteins, or other materials, but they were not alive since they had no metabolism and could not reproduce. In 2011, the first artificial cell membrane was announced and, by 2014, the first synthetic eukaryotic cell had been produced with working organelles and multiple internal chemical reactions. Other experiments replaced some parts of the essential cell machinery with artificially generated substitutes. For example, in 2010, a group created synthetic DNA and used it to replace the DNA in a strain of bacteria that continued to replicate. No one has yet attempted the artificial construction of a realistic multicellular organ or organism.

Researchers have long known how to do synthesis by mathematics. Still, applications have been severely limited by the available methods until sufficient computing power became available in the latter part of the 20th century. Now it is possible to investigate the behavior of theoretical systems of thousands or even millions of linked holons through numerical computer methods. What are the implications of such synthesis calculations?

There were hints of parts in relationships creating new properties in the writings of early philosophers. Aristotle wrote that "the whole is greater than the sum of its parts." The idea of the "atom" as the smallest part also began with Aristotle. Plotinus, the first to spell out the principles behind the Great Chain of Being, used the term "emanation" to describe how the Chain of created beings and matter originated from the "One," Plato's "Idea of the Good" or Aristotle's "God." The Great Chain was a "top-down" explanation for the universe, not involving complexity, but rather the element of "soul." Although Plato and Plotinus did not include the possibility of upward movement in the Chain, the Neoplatonic philosophers that followed eventually did allow for the development of new species and upward movement in the Chain as a goal for humans.

What about *de* novo synthesis, linking together parts or holons *not* derived from a reduction? Linking together holons at any level in the Catalog of the Universe in defined relationships, whether done experimentally or theoretically through mathematics and computing, is nothing less than the *creation* of something new, something more complex – the creation of a higher level in the Catalog or an addition to a higher level. The new creation will have unique properties not displayed by the holons at their lower level. Scientific synthesis is so significant that the process has a unique name: *emergence*, with the results being *emergent* properties.

The first person known to use the term "emergence" in the above sense was George Henry Lewes. In an 1874 book, Lewes describes how new properties "emerge" when molecules combine in chemical reactions:

> "We do not suppose that when what is called the physical motions of molecules are grouped into what is called chemical actions, and surprisingly novel phenomena emerge, there has been anything essentially superadded to the primitive molecules and their forces. Nor do biologists now suppose that when physical and chemical actions are specially grouped and vital phenomena emerge, anything essential has been superadded to the primitive threads of

> objective existence. The chemical phenomenon is new, the vital phenomenon is new; but the novelty is one of special grouping of the old material and the old energy." [1]

Lewes was only partially correct, however. When the "primitive molecules" react, something new *is* added. That something is the *information* describing *how* the molecules interact. In chemical reactions, the linking forces are interactions between the electron clouds of charged atoms or the higher-order electrical interactions between neutral atoms. Higher in the Catalog, interactions between cells, organs, or species can be more complex, involving electrical, chemical, liquid flow, and mechanical forces. While the "information" of interaction is not matter added to the interacting system, it adds to the interacting elements' total energy. The added energy is a measurable quantity.

Scientists discussing the concept of emergence have divided themselves into two groups, emphasizing *strong* or *weak* emergence. Here is how Philip Clayton describes the categories:

> "Strong emergentists maintain that evolution in the cosmos produces new, ontologically distinct levels, which are characterized by their own distinct laws or regularities and causal forces. By contrast, weak emergentists insist that, as new patterns emerge, the fundamental causal processes remain those of physics." [2]

Sean Carroll gives a helpful illustration of emergence and what weak and strong imply. [3] Think about the properties of the air in your room. Normal air is a multicomponent gas, and its properties can be described by pressure and density and the partial pressures of nitrogen (78%), oxygen (21%), and argon (1%) that make up normal air. But we now also know that nitrogen and oxygen are atoms in molecular form N_2, O_2, and Ar. We could describe air in the room by listing each of these molecules and their position in the room. The first is a macro description, the second a micro description (and an exceedingly long list!) The macro description is the weak emergent behavior of all the molecules, in theory, derivable from all the molecules' properties in the room and their interactions by a statistical calculation. Usually, we are happy to accept the more straightforward emergent description. If there was strong emergence, the molecules' interactions would be changing the fundamental physics of the molecules and atoms. There is no evidence of that happening.

Long before the Catalog of the Universe's unification, scientists in fields such as geology, biology, or physiology developed empirical rules or laws for the systems they studied, based strictly on their observations and not on fundamental laws or sub-holons. Scientists now know much more at all the Catalog levels, but such empirical rules and laws remain helpful, as Clayton notes. They need not be discarded. There is, reasonably, no demand from the scientific community that such higher-level laws now must be derived from fundamental principles. However, there is value in deriving these laws or regularities from one or two levels down the Catalog by synthesis modeling, finding causal links, and gaining as much insight as possible for a level. In physiology, this activity is now called "translational physiology," where clinical issues are linked to genetic or other molecular mechanisms multiple levels down the Catalog.

The number of possible holons at a level is called the "span" at that level, and the possible span increases dramatically at higher levels of the Catalog. The actual number of stable new holons may not be significant. The Standard Model of Particle Physics at the bottom of the known Catalog presently includes 17 particles. Linking combinations of these 17 to produce different atom types results in a Periodic Table of 118, with only the first 94 occurring naturally while numbers 95-118 have been created artificially. By the next level up – molecules – the number of possibilities is much, much larger. The count within three levels thus goes from 17 to 118, and then innumerable!

As an example of the enormous span of molecules, the carbon atom is recognized as the basis of all life on earth, animal and plant, combined with other atom types. How many different molecules could be constructed out of carbon? Twenty carbon atoms can be put together in 366,319 different molecular structures. Increase the number of carbon atoms to 40, and there are 62 trillion possibilities. [4] Add oxygen atoms, and the number of possibilities is even larger! Twenty carbon atoms is a tiny number compared with many carbon-based biological molecules. One strand of human DNA contains approximately 30 billion carbon atoms. The number of possible molecules that could be constructed with 30 billion carbon atoms is beyond astronomical! Further up the Catalog, the theoretical number of possibilities is even higher.

How many levels there are in the Catalog of the Universe is called its "depth." My illustration in Chapter 2 shows 23 levels, but I did not intend it as a complete Catalog. In a book entitled *Emergence: From Chaos to Order* [5] Holland listed eight levels. Harold Morowitz wrote about the history of development in 28 levels in *The Emergence of Everything: How the World Became Complex.* [6] Ridley's book with a similar title, *The Evolution of Everything: How Ideas Emerge* [7] considered mostly human interactions in 16 areas. A complete Catalog of the Universe, including all known genetic information, all known ecologies, and cosmic structures, would be a *vast* and detailed graphic!

If the ancient philosophers had any idea of the richness of possibility in our universe, they never would have expected that the Great Chain of Being contained everything possible to be created!

Why is there such an explosion of possibilities as complexity increases? Consider LEGOs, a system of interlinking multicolored plastic bricks, first made in 1949 with the current brick design introduced in 1958. Nominally a construction toy for children, LEGOs created buildings, towers, and walls, but that was just the beginning. There followed an explosion of kits to build all manner of wheeled and flying vehicles, spacecraft, construction machinery, and specialized buildings to make cities. For adults, there have been large-scale models of vehicles and museum-sized pieces of art. A graphic software program is available to design Lego-based objects before assembly. In 1999, designers added motors and electronics to create robot kits. Why such a wide range of objects created? Because the blocks can be linked to one another in so many ways. Just six bricks, each having 2 × 4 linking studs, can be combined in 915,103,765 ways! [8]

In mathematical terms, more complexity means more state variables necessary to describe sub-holons' states. Additional information is required to describe the interactions. Both factors contribute to the number of parameters needed to specify the system. The range of possibilities for the synthesized system is the *product* (not the sum) of the possibilities for all the system's parameters, harking back to the idea of "parameter space" described earlier. Each parameter contributes a new dimension to the parameter space.

The ancients may have known some fossils, but it was not until the 17th century that scientists recognized fossils as originating from earlier forms of life. Since that time, fossils of many forms of life have been found and identified, making it clear that the living species on the earth have changed with time. Biologists estimate that the present number of living species on the planet is around 10 million, of which about 1.2 million have been observed directly. But the total number of species that have lived on earth is estimated at 100 times more. In other words, more than 99% of the species that have lived on earth are now extinct. Species continue to disappear, and new species appear, so the living count is not constant. Even these numbers, while large, are small compared with any estimated theoretical number of possible species. The recognition starting in the 17th century CE of species changes was just one part of a general recognition of changes in the natural order, including in geology, cosmology, and life forms.

At the level of molecules and above in the Catalog (i.e., most of it), we, therefore, must recognize how little we know about what can be created within the laws we know about! If the possibilities in the Catalog of the Universe are so unlimited, can we make a definite judgment about any event, classifying it as natural or not-natural? Shouldn't we be able to tell if a physical law is broken?

Suppose someone asks you to invest in a company building a new machine the size of a small suitcase that produces electrical energy for years with no fuel to buy. Just think, you can disconnect from the grid and have no more electrical bills! The device has no tank to fill with gasoline or liquid hydrogen, no place to insert wood or coal to burn, and no solar cells or rotating blades to put on your roof. Would you invest in this company? A hundred years ago, your physicist friend would say, "don't invest; this is a trick because there has to be a source for this energy." Today, such a device is not only possible but in regular use: generating electricity from radioactive decay. Generators of this type have been used in spacecraft since the 1960s, although you cannot buy one for your home for various safety and technical reasons.

Scientists knew about energy conservation since the 18th century, but they did not know about radioactivity or how to convert mass to energy until much more recently. A judgment of any event or machine's natural possibility or impossibility depends on the current state of scientific and technical knowledge.

Not all theoretically possible combinations at any level are expected to exist. First, systems of interacting holons must obey physical laws, such as the conservation of energy. Second, any interacting system must have some stability in time – it must at least be stable long enough for humans to observe if we are to be interested and much more stable if it is to be part of some human technology

Why, then, do we live in this particular world with a specific tiny subset of what must be the possible range of life forms? It does leave a lot of space for the imagination of science fiction writers! At present, no one knows why we humans inhabit a planet with this particular set of branches, leaves, and connections in our Catalog of the Universe. The evidence is that our earth, its inhabitants, and the rest of the universe developed through many steps and processes (evolution).

If the universe's development from the Big Bang repeated, would the result be an entirely new universe with a different Catalog, or would the resulting earth be similar or even identical to ours? We have no way of rolling time back to replay the development sequence to check. Some scientists believe that because the evolution of the universe and life on earth involve random processes, the results would be entirely different. On the other hand, Simon Conway Morris and his co-authors describe the phenomenon of *convergence.* [9]. This word means the independent evolution of similar features in species of different periods or epochs in time. Convergent evolution creates *analogous structures* with similar forms or functions that were not present in those last common ancestor groups.

Examples of convergence abound. One is the camera eye, a curved layer of photosensitive cells with a lens in front to focus light. This sensor structure evolved separately by vertebrates – including humans and other mammals, and by the squid, octopus, and jellyfish! Another example of convergence is a seed dispersal method developed by plants: an attractive bait "cap" added to each seed. These caps attract insects who carry the seeds off to their distant homes, thereby dispersing the plant's seeds. This "trick" of seed dispersal evolved separately more than 100 times, ending up in 11,000 species. Other examples of convergence exist, down to the molecular level.

In each of these examples, the evolutionary solutions converge not because of random coincidences but because physical laws limit possible solutions. In the example of eyes, the properties of light limit possible eye architectures. Evolution found only one more light sensor structure in addition to the camera eye: the compound eye of the fly. This structure consists of an array of tiny light sensors on a curved surface. It is no coincidence that in the 21st century, radar antennas – a kind of "eye" using short-wavelength radio waves instead of light – are built using the same two designs: a single parabolic "lens" and an array of small antennas. Convergence occurs in these two cases because light and radar waves are both electromagnetic phenomena that follow the same laws.

A rerun of evolution on planet earth thus would probably produce a mixture of repeats and new solutions. In the next chapter, we will see that synthesis is the fundamental basis of creativity that populates the Catalog.

CHAPTER 15

Human Creativity

Creativity is intelligence having fun. *Albert Einstein*

Moving up a level in the Catalog of the Universe means selecting holons from one level and putting them in interacting relationships to construct something of higher complexity. This process is *synthesis*, and there is also another word to describe this process: *creating*. It is creating complexity, but in practice, we call the act simply Creating.

"Creative" is an adjective widely applied to capable and admired people. Artists create drawings, paintings, and sculptures. Writers create essays and books. Composers and musicians create music. Poets create poetry. Engineers create bridges, roads, and assembly lines. Architects, working with engineers and builders, create buildings. Comedians create laughter and fun. Dancers create beauty in form and motion. Scientists create understanding of natural phenomena. Philosophers create organized worlds of thought. Mathematicians create worlds of symbols and logic. Computer programmers create code. And the list could go on and on.

The June 2016 issue of the business magazine *Fast Company* featured their list of the 100 most creative people for the previous year. Many made the list because of extraordinary success in business and technology. For example, Jens Bergensten was #5 on the list as the lead designer of the online game Minecraft with 100 million players. #27 was Jennifer Lewis, who developed a technology that allows 3D printers to print electronic circuits. But not all people on the list were from the world of business or technology. #1 was Charles Arntzen, who developed a treatment for Ebola using the tobacco plant. Comedian Amy Poehler was #8 for finding multiple new ways to bring comedy to audiences. Position 41 was Vian Dakhil, a Member of Parliament in Iraq, for launching a worldwide crusade to save the Yazidi religious minority people who were surrounded and threatened by ISIS. Perhaps the most unusual was #65, tattoo artist Vinnie Myers who helps restore women's image by adding realistic nipples to surgically reconstructed breasts.

Each human creator works within a discipline where various building blocks (holons) exist: materials, objects, processes, symbols, or ideas. The creator attempts to find new ways to combine these building blocks to bring about unique beauty, new functionality, new ideas, or just satisfaction. Painters have a wide range of hues available, a means of arranging the colors on the canvas, and themes. The creative painter aims for an arrangement of patterns and colors that generates the desired response in viewers. Musicians strive for combinations of melodies, harmonies, rhythms, and tones that resonate and captivate listeners. Engineers accept the challenge of a task and work to create something that quickly and efficiently accomplishes the task. Chemists put different combinations of atoms together to create various molecules, looking for desired properties.

Is it correct to use the word "create" in the manner I have suggested? After all, the painter doesn't make the paints or canvas from nothing, and the chemist uses existing atoms in building molecules. It is a lesson from the Catalog of the Universe that creating occurs at all levels, not just before or at the Big Bang or when the first carbon atoms began to appear. In the era of science, we humans can now recognize the hierarchical structure of the universe. Each step up in complexity is a genuine act of creation, whether by unaided natural process or through human or other intelligent intervention.

It *is* essential to think about the Catalog's very roots: Where did the particles in the Standard Model and their interactions' laws come from? Are there more hierarchical levels below the Standard Model, or is there a lowest root level? We can trace the origin of atoms making up humans and the universe back to the Big Bang. But even that beginning had to occur in an environment of physical laws. Those laws' origin is beyond the scope of this book, although I will have a few things to say on the subject in the next chapter.

Are there limits to human creativity? As a child, I wondered if there was a limited number of books that could be written, paintings that could be painted, or musical pieces that could be composed. Just like the vast number of possible molecules based on carbon atoms, I know now that there is no need to worry that succeeding generations will be left with nothing new to create in the arts.

Consider an example from digital photography. Anyone who has bought and used a camera in recent decades knows that the detail in a digital picture, the number of picture elements or pixels that make up the picture, determines its detail. Each pixel can have a unique color determined by the camera sensor's capabilities and recording electronics and ultimately by the human eye. Even a modest camera today can record 10 million pixels for one picture. Estimates of the number of distinct colors recognizable by the human eye range from 2 to 100 million. As a conservative example, how many different images could be generated by an array of 1 million pixels, each capable of 1 million colors? The number of possible "pictures" is so large that it strains scientific notation, designed to make large numbers palatable: 1 followed by 6 million zeros!! For comparison, the number of atoms in the known universe has been estimated at 1, followed by "only" 80 zeros.

Our imaginary picture generator would generate every possible picture in the universe – everything in the past and every possible thing in the future! Of course, almost all the virtually infinite number of images possible will be of no interest and would simply be called visual noise. So, it is the photographer or computer artist's job to create combinations of pixels and colors that attract and hold our attention. Some of the successful combinations have come from recording an image from the real world, some from modified or imaginary images, and others have been abstract images that have no origin in real-world observation. A similar argument would hold in any field – visual arts, music, writing, construction of buildings or physical devices, or even scientific experiments. The number of possibilities is so large that humanity will never run out of new things to create.

Why are some people more creative than others? Psychologists and brain scientists have conducted many studies in search of clues to increased creativity. The field is still in its infancy, but one theme seems clear from the studies to date: Creativity increases with diversity and variety on several levels. Individuals who interact with multiple cultures or environments are more creative than individuals who live entirely within one uniform group. Also, teams are more creative when the team includes both sexes and individuals with different fields of training and experience. Even social diversity in a team is associated with increased creativity. Debaters and interviewees who will face an opponent with known differences are more creative in their interactions because the anticipation causes them to prepare better. Psychiatrist and neuroscientist Nancy Andreasen concluded from her studies that "creative people are better at recognizing relationships, making associations and connections, and seeing things in an original way – seeing things that others cannot see." [1] Talking to yourself or what psychologists call "inner speech" may also contribute to creativity, for example as a person talks through both sides of an issue internally. [2]

Successful creativity requires more than just picking out some holons – bricks, pixel colors, or notes – and putting them into some relationship. The crucial step is recognizing when the new linked grouping has value and appeal, for example, because of its beauty or because it offers a unique and valuable function. Numerous scientific and technical projects and companies have failed because no one appreciated a new creation's actual worth. Thomas Edison is famous as the inventor of sound recording. Twenty years before Edison's invention, Édouard-Léon Scott de Martinville filed a patent for a sound recording machine. He was a printer by trade but also studied the anatomy of the human ear and the art of steganography (concealing secret messages within non-secret text or pictures). His device funneled sound waves through a horn structure to vibrate a membrane with an attached stylus that wrote waves on a page darkened by lampblack. De Martinville named his device the "phonautograph." Why do we remember Edison and not the phonautograph invented twenty years earlier? Because de Martinville never thought of adding a playback device to his invention. You could only look at the squiggles on the black paper recording but not listen to them, and the eye is not capable of translating the squiggles back into audible sound. It seems evident that playback must go with recording, but it was a blind spot for de Martinville and spoiled his creative act.

The famous physicist Richard Feynman said something that gives us another clue about creating: "What I cannot create, I do not understand." The creator must understand the materials, words, sounds, or ideas at hand to form a new creation. That understanding usually comes from long hours of experience, trial, and error. New musicians or young athletes eyeing careers as a soloist or a world-class athlete learn that it takes 10,000 hours of practice to achieve a professional level [3], although this number is controversial. [4]

Some creative acts, especially in science and technology, can change human thinking, culture, or daily life. Examples would be Kepler's laws of planetary motion, Gutenberg's printing press, or the transistor at Bell Labs. What is there about the people and conditions that make such remarkable steps possible? In his book The Act of Creation[5], Arthur Koestler has extensively analyzed such cases and has identified some necessary conditions. The potential creator must, as Koestler puts it, be "ripe" for the new insight. That is, the potential creator must have been thinking about the problem to be solved and consciously or unconsciously trying out possible solutions. The creative act itself is usually triggered when two separate areas of thought (Koestler's term is "matrices") intersect, when some idea from a field generally unrelated to the desired solution's field has an unexpected overlap.

For example, Gutenberg had been thinking about how to make multiple copies of the Bible. The standard method of making "books" at the time was to carve all the letters for a page on the surface of a block of wood, rub some ink on the letters, lay a sheet of damp paper on the inked block, hand rub the back of the paper until it was smooth, then peel the paper off the wood block. Paper could only be printed on one side with this process, so to make a book page, two "printed" pages would be glued back-to-back. The process was slow and tedious both because a whole new block had to be carved for each page and because the hand-rubbing process was slow. Koestler identified this block-printing process as "Matrix 1."

Matrix 2 for Gutenberg was the process of making coins. This process began with a small steel rod with the desired relief pattern carved onto its end. This rod was then hammered against another piece of steel, creating a mold or "stamp" for the selected coin. Finally, a gold disc was placed against the mold and given a sharp blow, forcing the soft gold into the mold and creating the coin's desired image. The hard mold could be used repeatedly to make many coins before the relief pattern would wear out. It occurred to Gutenberg that he could use a similar process to create individual letters that could be arranged in rows and blocks as a page of text, then taken apart and rearranged for another page of text. He cast his letters in lead so he could make many copies of each letter. The reusable letters then made it much easier and faster to create the "block pattern" for each page.

For Gutenberg, there was a second creative insight stemming from a Matrix 3. After he participated in a wine harvest, he noticed how the winepress's great power drove the juice out of the grapes and realized that a press could also force paper against his new reusable type blocks.

Thus, the printing press was born from two creative acts, stimulated by the intersections of three unrelated fields. In his book, Koestler recognized creativity in many areas as following the same pattern: humor, philosophical ideas, all forms of writing, visual arts, and human development, to mention just a few.

Koestler's insights into creativity can illuminate educators and individuals in the 21st century who want to inspire creativity in others or their own lives. The lesson is clear: participate, encourage participation, or observe activities in as many different areas as possible. There is no way to predict which areas might overlap to spark creativity, so multiply the available matrices to increase the chances of a productive intersection!

What does a multidisciplinary team look like? My experience with an early example was the cardiac electrophysiology group at Duke. Dr. Spach, one of the two team leaders, was a practicing pediatric cardiologist, spending 20% of his time working with patients and the other 80% in his basic research laboratory. Dr. Johnson, the other team leader, had a medical degree from the University of Sheffield, UK, but was a full-time researcher and teacher and did not see patients. Other members of the team included: a specialist in growing heart tissue in culture, another practicing cardiologist, a specialist in the microanatomy of heart cells and tissue, a biomedical engineer specializing in mathematical models, an electronic/mechanical engineer who could build any needed instrumentation in his well-equipped shop, and usually a couple of graduate students – most often with a background in physical science or engineering. I added my experience in computing and mathematics to this team, along with my recently learned physiology.

This diverse group met several times a week for two or three hours to work through ideas for the group's research. These sessions continued through all the years I was at Duke. They were unlike anything else I have experienced, before or after. You might be picturing the typical committee meeting where the higher-ranking attendees dominate the exchanges, and the junior participants try to think of something cogent to say. There was no hierarchy in the Duke sessions; anyone who had a question or idea to contribute, whether graduate student or lab chief, was given equal respect, and first names were appropriate for everyone. Group members used the blackboard liberally to construct lists, process diagrams, or mathematical equations. The level of concentration by everyone was intense; the volume occasionally rising almost to the level of shouting! Surprisingly, this intensity contained no anger or personal animosity. Even a graduate student's poorly thought-out idea was not treated harshly. When each session ended, all the participants exited the room chatting amiably, often in smaller groups to share lunch. I usually joined Dr. Johnson because of our mutual interest in constructing theories, careful writing of our research papers, and music. We spent the "lunch" hour together each day jogging on the wooded track around the beautiful Duke golf course, talking about things as diverse as family issues, music, philosophy, and religion.

I might have come to believe that this was the way research was done in all productive labs at this time, but one incident showed me that this was not the case. We had a visit one day from a big-name researcher in our field from a very well-respected university. He sat with us through the morning discussion, saying relatively little, though he was prominent in the area and usually outspoken. When the session ended, and we all filed out of the room, he shook his head, saying, "How do you stand this intensity every day!"

This physiology research group was productive because the participants could interact without ego clashes and the group's breadth of experience. The latest instrumentation was designed and built to order within the group. The group developed a detailed knowledge of heart cell and tissue anatomy. Participants understood cellular mechanisms like membrane transporters and pumps and clinical experience with heart disease so real patient issues could guide the research. And there was experience in constructing and testing mathematical theories for the observed electrical behaviors of heart cells and tissues.

My participation in this Duke research group ended 30 years ago, and by now, multidisciplinary research groups are not as unique as they were then. Many groups have discovered the benefits of diversity in a team. It is much more common to find multiple authors on papers on medical and biological subjects. However, I wonder how many of these current groups achieve the level of egality and cooperation managed by the earlier Duke group!

In the 30 years since I left the Duke faculty, many universities in the US and around the world have created departments and degrees in computer-based mathematical modeling and simulation for biomedical systems. Multidisciplinary research groups have also become more common as both the experimental techniques and theories for biological and medical systems have become more complex, involving more related fields.

In the next chapter, I will introduce the method used to study the behavior of groups of holons from the Catalog, whether human, other life forms, or inorganic matter.

CHAPTER 16

Living Together

> Great things are done by a series of small things brought together. *Vincent van Gogh*

In between the Catalog of the Universe layer made up of living things and the layers of the cosmos' bodies lies a layer of particular interest not just to scientists but to all humans. Living things are all holons in the hierarchy terminology and can be combined to create the more complex holons in this particular layer. At the level of living things in the Catalog, the span is vast, including every species of plants and animals in the Phylogenetic Tree of Life shown in Figure 7.1, including humans – millions of species. What can we learn about the living things groups by considering them as holons in a hierarchical level in the Catalog of the Universe?

The most general new holons above the level of individual species are ecologies -- associations of communicating and interdependent animals and plants, their geological environments, local atmospheric conditions, and even the earth's position in the solar system (via the seasons). While some ancient Greek philosophers were interested in natural phenomena, *essentialism* governed their thinking: the belief that everything in nature was static, continuing as God created it. There was little interest in interactions. This philosophy held sway until the 16th century when realization began to dawn that there was change in nature.

The first ecology concepts originated in the 18th century with microscopist Antoni van Leeuwenhoek and botanist Richard Bradley, who wrote about food chains, population regulation, and productivity. The German biologist Ernst Haeckel coined the term "ecology" in 1866. There was great interest in observing nature and recording observations in detail through the 18th and 19th centuries, including Darwin's work, but the available analytic mathematics severely limited ecology theories.

Thomas Malthus began theories of population dynamics with his 1798 *Essay on the Principle of Population.* Malthus argued that populations grow geometrically (i.e., with increasing speed) while resources remain constant or only grow arithmetically (steadily). It took another 40 years before Verhuis (1838) expressed Malthus' concepts as a mathematical equation called the *logistic* equation. This equation described the population of only one species, however.

Another 90 years went by before Alfred J. Lotka (1925) and Vito Volterra (1928) derived, from the chemical principle of mass action, mathematical equations for the relationship between the populations of two species, predator and prey. [1] The Lotka-Volterra model received some enhancements in the subsequent years. Still, it remains a teaching example for biology and ecology students, applied to foxes and rabbits, for example. Solving the Lotka-Volterra equations shows a continuing oscillatory behavior of the two populations. The number of rabbits goes down as the foxes eat them, accompanied by a decrease in foxes when fewer rabbits are available to eat. With less foxes, the number of rabbits begins to increase, and the cycle starts over again. Students studying these equations learn, for example, that changing the rate at which rabbits reproduce changes the frequency of oscillation of the two populations but causes little change in the size of the population swings.

An ecology with two interacting species is a very simple ecological model, but that was about as complex as analytic mathematics could handle. As described earlier, digital computers have lifted that limitation beginning in the middle of the 20th century. The possibility of including more than two interacting holons in ecological models made them more realistic and valuable to society, so interest in the field has grown sharply. Researchers study these larger ecological models as *networks.*

A network is a group of *nodes* (individual holons) connected by *links* representing relationships. The nodes in a network can be all the same type, or they can be different types, possibly even from multiple levels in the Catalog. Similarly, the links in a network can all represent the same kind or different relationships.

Nodes in a network can be a molecule in a biochemical reaction, a person in a social network, or a sun in a galaxy. In an ecological network, such as Figure 16.1, each node is a species from a different place in the Catalog. Multiple holons of the same type can also form a network, for example, the groups of fireflies of Southeast Asia who flash in synchrony [2] or the fish in a school.

Links between nodes in a network can represent any interaction type. The fireflies detect each other's light. Each star in a galaxy feels the gravitational pull of all the other stars. In a forest, fungi link tree roots together to help the trees share resources. [3,4] In an organ like a heart, multiple cells coordinate their work by exchanging electrical or chemical signals. Scientists are still being surprised by newly discovered communication paths, for example, how intestinal tract bacteria help regulate blood pressure in the humans they inhabit. [5] It may be helpful to think of the communication between holons as a transfer of information, but information is not transferred by itself, only by a physical process.

Figure 16.1 shows an example of a combined ecological-social network in Madagascar. The "actors" are clans that interact with and manage the forest patches. The network has three types of interactions: forest-to-forest, clan-to-clan, and clan-to-forest.

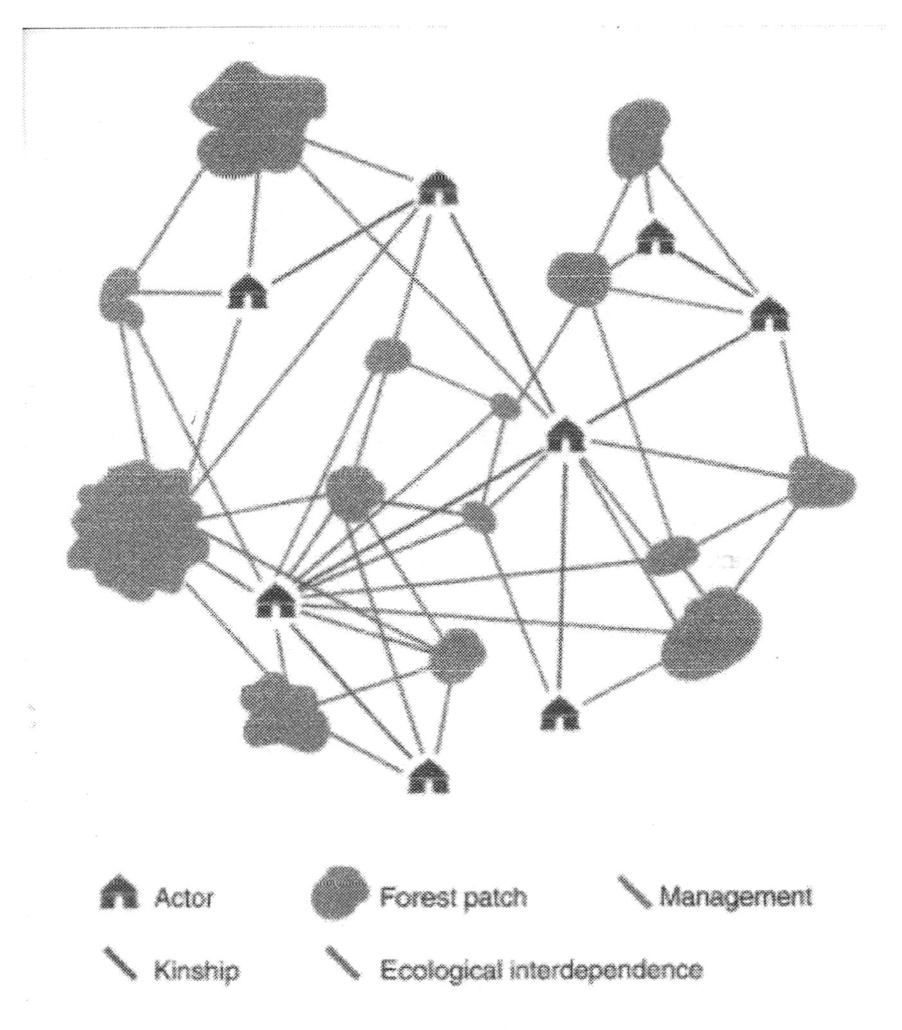

Figure 16.1

A network of forests and clans in Madagascar. [6]

Ecological networks frequently have more participants than is first suspected. For example, African dryland studies focused on the acacia trees and bunch grasses and grazing and predatory animals. When satellite photos became available, it was clear that termite mounds were more numerous and evenly spaced than was expected – just far enough apart, so the termites avoided territorial battles. More extensive satellite monitoring showed that these mounds were hotspots for plant growth. [7,8] The tiny termite with a very simple nervous system that does not even include the design of the mounds the colony builds is thus a participant in the ecology along with the vastly larger animals.

Scientists are now working to build artificial ecologies, for example, communities of different types of microbes. [9,10] Such mixed populations can simultaneously perform otherwise incompatible functions, and they resist loss of function from environmental perturbation. In one example, researchers put together two strains of *E. coli* bacteria with different genetic structures, one being an "activator" and the other a "repressor." The two strains communicated by exchanging signaling molecules. The result was an oscillator or repetitive cycling – a behavior not possible with a single strain. Synthetic ecology may thus be a promising approach for developing robust, stable, biotechnological processes, such as converting cellulose to biofuel.

In addition to ecologies, networks model many kinds of processes. Here are some examples of types of networks:

- Networks with nodes that can make decisions or change values
    - Chemical reactions
    - Multicellular tissue
    - Games – from chess to soccer
    - Competing businesses that make similar products
    - Bidding in an auction
    - Choosing a route through a transportation network (e.g., cities and connecting roads)
    - Deciding on a policy in international relations
    - Communications technologies
- Networks with nodes where no one is overtly making decisions
    - Evolutionary biology
    - Success or failure of new cultural practices and conventions

Scientists frequently construct a network model for a specific process. For example, a network might follow nitrogen through a cycle from the ground into plants consumed by animals eaten by humans, whose waste is processed into fertilizer, eventually ending up in the ground again. The study of such a network can help maintain a habitable environment.

A valuable network organization quality for ecological and other networks is modularity, grouping nodes together in clusters with fewer connections between the clusters (Figure 16.2). When the network is modular, a perturbation starting in one node is less likely to spread throughout the web because of the modules' partial isolation. [11,12] An example of this would be the spread of a disease through a population. For a given population size, the condition would be less likely to spread throughout that number of people if they clustered in several small communities, compared with the same number living together in one large city.

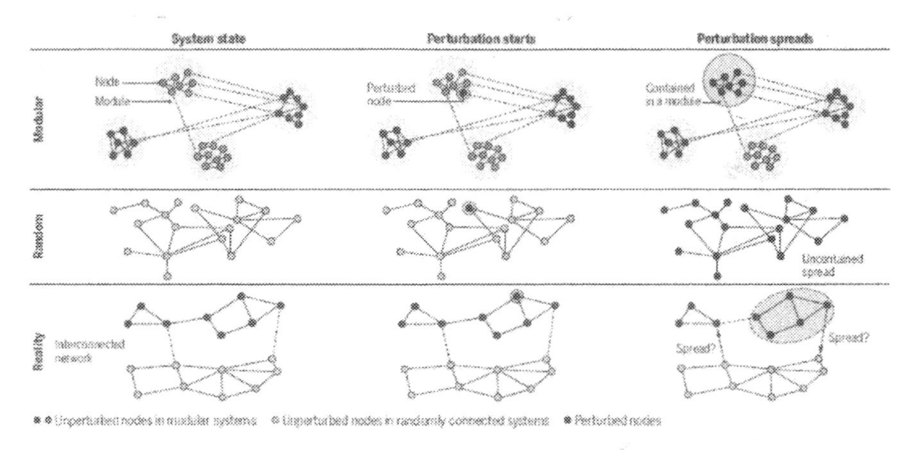

Figure 16.2
Modular networks. [11]

The opposite of the modular network is the swarm: many identical nodes with tight coupling between them. Examples of this behavior are swarms of fish swimming together, clouds of locusts, flocks of sheep, and crowds of humans. [13] One of the most remarkable swarms to watch is a large flock of starlings called a murmuration [14] where the swarm wheels back and forth through the sky, the shape of the mass constantly changing with graceful curves. Biologists believe that fish swarms or schools provide protection, and the same may be true for the starlings, but they seem to fly in the same rolling swarms even when no predator is near. Recently, engineers imitated a bird swarm with mechanical devices. [15] They built and flew a swarm of more than 100 tiny robotic drones, dropping them at a location from a jet fighter aircraft. These little crafts were each autonomous in their flight control, and programmers could direct the swarm to move to a different location or shape by a human controller or by an internal program. Drone swarms now generate complex shapes in nighttime displays.

Because networks arise in so many different situations, a substantial body of mathematical theory applies to networks. [16] Game Theory can describe a network's behavior, for example, the amount of nitrogen in each step of the cycle described above or how the nitrogen flow through the network would change in response to a change in one of the network's nodes. Game Theory is complex mathematically, but one network property can be described in simple terms: If one node in a network changes its value or interactions, all the other nodes in the network are – at least in theory – affected as well.

A second type of theory for networks is Graph Theory. Graph Theory is primarily concerned with networks' structure made up of *vertices* (nodes) and *edges* (links). Edges are simple interactions that do not have names or complex properties like links in a public network, only the possibility of a direction. Although simpler in structure than public networks, there are many applications where just the existence of an edge is sufficient detail. Two examples would be decision networks and linguistics. Graph theory includes methods for transforming graphs to simplify them or make them more suitable for technical applications.

The histories of game and graph theory began with a paper written by Leonard Euler on the *Seven Bridges of Konigsberg* and published in 1736. Königsberg (now Kaliningrad, Russia) was situated on both sides of the Pregel River and included two islands connected to each other and the two mainland portions of the city by seven bridges. The challenge was to devise a walk through the city that would cross each bridge once and only once. Also, solutions involving either reaching an island or mainland bank other than via one of the bridges or accessing any bridge without crossing to its other end are explicitly unacceptable. Euler proved that the Königsberg problem has no exact solution. But he had no suitable mathematical technique that would provide a general proof for problems of this type. This problem came to be known as the "traveling salesman problem:" Given a list of cities and the roads between them, what is the shortest route a salesman can take to visit each city and return to the starting city?

Nineteenth-century papers developed Euler's ideas into general mathematical theory. But it was not until 1972 that Richard Karp showed that no algorithm for the traveling salesman problem was guaranteed to find the shortest solution. [17,18] Work continued on the problem, however, because of its practical importance in planning delivery routes. In 1976, Nicos Christofides developed an algorithm that produces routes guaranteed to be at most 50 percent longer than the (unknown) shortest route. [18]

The standard traveling salesman problem assumes that travel between any pair of cities costs the same in either direction. This assumption is not always valid. For example, airline flights might cost less in one direction than the other because of headwinds. In 2017, Ola Svensson, Jakub Tarnawski, and László Végh produced a new approximate algorithm that works for both the symmetrical and the asymmetrical traveling salesman problems. [18]

In summary, when studying the natural universe, we can use two kinds of organizational structures: hierarchies and networks. The former helps us organize the *types* of holons existing and possible. The latter shows up when we attempt to understand *functional systems* made up of different kinds of holons, like the solar system or ecologies.

CHAPTER 17

Humans and Human Groups in the Catalog

> Relationships are all there is. Everything in the universe only exists because it is in relationship to everything else. Nothing exists in isolation. We have to stop pretending that we are individuals that can go it alone. *Margaret Wheatley.*

The span of new holons made possible by linking together individual living holons is vast beyond comprehension. In this chapter, I wish to focus on one fraction of that span: human groups and organizations.

In the spirit of scientific reduction, the human body has been dissected and analyzed in great detail for medical purposes and pure scientific understanding. The body's organ systems are well known, and there is extensive knowledge about the cells making up the organs of the body and the biochemistry in those cells. Nevertheless, there is still much to be learned about the human body and what can go wrong. In 2016, there were 567,890 research papers published in medicine and another 266,385 in biochemistry, genetics, and molecular biology. Also, both counts continue to rise year after year. The brain remains the most complex and mysterious organ of the body, even though its basic unit – the neuron – is one of the best-understood cell types in the body.

As anyone knows who has tried to locate a friend in a large crowd, the span of human holons is vast. There are many variations in body size and shape, skin color and texture, hair color and texture, amount, and style, not to mention clothing and movement characteristics.

As a component of a more complex system, the human as a holon is unsurpassed in the versatility of possible links with like holons. Basic mechanisms include facial gestures, body movements – especially hand gestures, touching -- vocalized sounds, tastes, and odors. Besides, there is a pattern-recognizing brain to identify and remember sensory patterns.

HUMAN GROUPS

Yuval Noah Harari, in *Sapiens* [1], his history of the human species, describes how human communications have changed over millennia and have determined the structures of human groups and human history. We know little of the lives of early hunter-gatherer bands because they left little physical evidence, but they were probably organized as families and tribes. Once agriculture began, human activities became specialized. Humans could live together in towns and cities, necessitating communications to organize life in these groups, such as keeping records of property ownership and distributing crops.

Humans have constructed complex languages, extensive libraries of symbols they can exchange via the basic sense mechanisms better than any other known living creature. Technologies have extended the range of symbol communication to cover the earth and extend out into space. Most English speakers use about 5,000 different words in everyday communications, but a person may have a vocabulary of 20-35,000 words. A college-educated person could know twice that many or more. As far as we know, humans are unique in their ability to formulate and communicate abstract ideas expressed through languages.

Other human activities are also varieties of communication, although not typically recognized as such. An example is money: a means of communicating value. Early money consisted of objects of accepted value in society, such as gold or shell pieces. In recent centuries, the things of value exchange became symbolic only with no inherent value. Examples are coins of common metals and paper money. Trust in the system is required to maintain the accepted values of the physical symbols. In our era, electronic bits represent value in computers, and we exchange them over communications networks. Yet another trust-based form of value communication is the idea of credit. As Harari explains [2], symbolic value and credit make many everyday human activities possible, from large-scale manufacturing, agriculture, and housing to the governing of large numbers of people.

Here are a few more categories of human group activities dependent on communications and designed to make life more pleasant and interesting:

1. Education	5. Scientific research	9. History
2. Literature	6. Entertainment	10. Science
3. Philosophy	7. Sports	11. Medicine
4. Visual arts	8. Music	12. Social Media

Each of these requires a specialized form of human interaction with its language and specialists. I did not intend this list to be complete, and you could probably add to it.

Like every other holon in the Catalog of the Universe, the human-group holons owe their existence and characteristics to relationships. Take one of these group holons apart, and – just like you would vanish if I disassembled you into clouds of protons, neutrons, and electrons – families, education, literature, and all the other items in the list above would cease to exist. What you would have left would be a large number of humans living isolated lives in separate caves!

The role of individual behaviors in human groups' success has been discussed by philosophers beginning more than 2000 years ago. In his writings, Plato had much to say about justice, for example, and the Greek word he used for this concept was "Dikaisyne." [3] The meaning of this Greek word includes not only the current meaning of the English word "justice" but also hints of "morality" and "righteousness." Plato was very dissatisfied with the degenerating conditions in Athens and wrote about justice in the *Republic*.

There were many justice theories in Plato's day, and Plato rejected most if not all of them. The traditional view was only proper conduct. Cephalus gave a more specific definition: to speak the truth and pay one's debt. Another description was that "justice is doing good to friends and harm to enemies." Plato rejected this traditional formula as not working in many cases. For example, suppose that an assumed friend was really an enemy; then what behavior is called for?

Thrasymachus had a more radical view of justice as "the interest of the stronger," i.e., might is right. Every man acts for himself and tries to get what he can. For a state, the Government is the strongest and takes what it wants for itself. Thrasymachus wrote

> "An unjust is superior to a just in character and intelligence. Injustice is a source of strength. Injustice brings happiness." [3]

In response to Thrasymachus, Plato answered

> "Just men are superior in character and intelligence and are more effective in action. As injustice implies ignorance, stupidity, and badness, it cannot be superior in character and intelligence. A just man is wiser because he acknowledges the principle of limit.
>
> Unlimited self-assertion is not a source of strength for any group organized for common purpose. Unlimited desire and claims lead to conflicts." [3]

In our scientific era, sociologists' work is to study people's behaviors in various group activities and attempt to explain each group's behavior in terms of the individual people involved, their characteristics, and the nature of their interactions. Even without reading the sociology literature, it is possible to predict that research in this area will be challenging. In hierarchical terms, both the individual holons (people) and their interactions are incredibly complex. Also, control of any experimental situation involving human behavior is problematic because moral issues like privacy arise. Researchers from fields much closer to the bottom of the Catalog of the Universe sometimes speak disparagingly of the "soft sciences" such as human sociology high in the Catalog. However, an understanding of the Catalog should generate respect for any well-designed research at these higher levels.

Just because there are links between humans, there is no guarantee that the results will be constructive. Individuals can choose harmful or destructive interactions for personal benefit or satisfaction, which harm the group or individual group members. Such a negative interaction is like a deadly virus in an interacting human group. It only requires a small percentage of cheaters, for example, to cause significant damage to a group's efforts because the few stimulate others to "even the score," spreading the damage. Sociological studies of corruption demonstrate this spreading effect.

> "Corruption, big or small, impedes the socioeconomic development of nations. It affects economic activities, weakens institutions, interferes with democracy and erodes the public's trust in government officials, politicians and their neighbors . . . Troublingly, our studies suggest that mere exposure to corruption is corrupting. Unless preventive measures are taken, dishonesty can spread stealthily and uninvited from person to person like a disease, eroding social norms and ethics—and once a culture of cheating and lying becomes entrenched, it can be difficult to dislodge." [4]

The researchers found that the tendency of individuals to cheat varies little from country to country. The level of corruption in any country varies instead with social norms and legal enforcement in the culture.

Hate groups certainly qualify as promoting negative human interactions. The Southern Poverty Law Center keeps track of the number of hate groups in the United States. [5] Their tally rose from 457 in 1999 to a peak of 1018 in 2011. The count dropped to 784 in 2014 and has been growing since, with 954 active groups in 2017. Who are these hate groups? The list includes the Ku Klux Klan, Neo-Nazis, White Nationalists, racist Skinheads, Black Nationalists, Anti-LGBTs, Anti-Muslims, and others. It seems that no racial, ethnic, or religious group is exempt from hating or being hated.

A MORAL CODE

Any grouping of humans, whether family, tribe, or nation, needs a moral code, a code that specifies the positive and negative or acceptable and unacceptable ways of relating within the group. Where does the moral code come from? Only two sources are known: Empiricism and Transcendentalism. Empiricism means that the moral code ultimately comes from human experience. In Transcendentalism, the moral code comes from outside human experience and physical reality, i.e., from an external supernatural source (god); I will discuss that type of authority in the next chapter.

Here is how evolutionary biologist Edward O. Wilson describes the empiricist approach:

> "The individual is seen as predisposed biologically to make certain choices. Through cultural evolution some of the choices are hardened into precepts, then laws, and, if the predisposition or coercion is strong enough, into a belief in the command of God or the natural order of the universe. The general empiricist principle takes this form: *Strong innate feeling and historical experience cause certain actions to be preferred; we have experienced them, and have weighed their consequences, and agree to conform with codes that express them. Let us take an oath upon the codes, invest our personal honor in them, and suffer punishment for their violation.* The empiricist view concedes that moral codes are devised to conform to some drives of human nature and to suppress others. *Ought* is the translation not of human nature but of the public will, which can be made increasingly wise and stable through an understanding of the needs and pitfalls of human nature." [6]

Philosopher Sam Harris has stated that "Science can answer moral questions." He points out that we humans are more concerned with what happens to other humans and to a somewhat lesser degree to animals than our level of concern about what happens to rocks. The difference is correlated with our understanding of biological complexity, expressed in this book as levels in the Catalog of the Universe. The higher complexity of humans implies a more fantastic range of experience than that of rocks. Harris says he has seen ". . . no notion, no version of human morality and values that is not at some point reducible to a statement about conscious experience." [7] For example, we should, says Harris, be able to look at the wide range of women's treatments in the various cultures and religions of the world and make some moral judgments about the extremes in this range, extremes we all know from experience to be painful or exploitive.

Author and philosopher Rebecca Goldstein suggests that philosophy should provide some basis for morality:

> "A common claim against the enlightenment has always been, and continues to be, that reason can provide no basis for morality. To anyone familiar with the long history of moral philosophy, this claim sounds as astonishingly uninformed as the assertion that science has provided us no basis for believing there are no laws of nature." [8]

Humans, other primates, and some other species display cooperative activities. Such cooperation seems to be an evolutionary anomaly:

> "In the hardscrabble competition for food, territory, and mates, why would one individual go out of its way to help another? Nevertheless, the animal world has plentiful examples of cooperation that seem to be hard-wired: bees that collect pollen for the whole hive, mole rats that build elaborate tunnels used by other group members, and meerkats that risk their lives to guard a common nest." [9]

Biologists first explained cooperation by a kin selection theory, sometimes called "inclusive fitness altruism." Helping one's relatives would increase the chances of one's genes passed on through them. Hamilton [10,11] gave the first explanation for altruistic behavior in 1964 in a simple equation that came to be known as Hamilton's Rule: $br > c$, where b is the benefit from the altruist, c is the cost paid by the altruist to deliver the benefit, and r is the relatedness or degree of kinship between altruist and recipient. In 1971, Robert Trivers published a paper that went beyond kin selection and started a decades-long discussion of "reciprocal altruism." [12]

Darwin recognized that humans were different from animals and wrote in 1871:

> "I fully subscribe to the judgment of those writers who maintain that of all the differences between man and the lower animals, the moral sense or conscience is by far the most important." [13]

Mayr echoed Darwin's statement in writing about a philosophy of biology:

> "The capacity for ethical behavior thus is closely correlated with the evolution of other characteristically human capacities. The difference between an animal, which acts instinctively, and a human being, who has the capacity for making choices, is the line of demarcation for ethics. . . .Since a person can predict the outcome of his or her actions, he [or she] is fully responsible for the ethical evaluation of the result. Human beings have the capacity to make such judgments because of the reasoning power provided by the evolving human brain. *The shift from an instinctive altruism based on inclusive fitness to an ethics based on decision making was perhaps the most important step in humanization."* [14]

Trivers wrote that "Each individual human is seen as possessing altruistic and cheating tendencies, the expression of which is sensitive to developmental variables that were selected to set the tendencies at a balance appropriate to the local social and ecological environment."

It is a common myth that humans are selfish, aggressive, and quick to panic by nature. Dutch biologist Frans de Waal calls this a *veneer theory*, the idea that civilization is a thin veneer that will crack at the slightest provocation. In his book Humankind, Rutger Bregman argues the opposite, that when disaster strikes, humans show their best altruistic side [15]. He quotes the example of Hurricane Katrina devastating New Orleans in August 2005. Eighty percent of homes flooded, and over 1800 people died. In the first week, the newspapers reported gangs, lootings, and a sniper aiming at rescue helicopters. Twenty-five thousand people were packed together in the Superdome, and it was reported that two infants had been killed and a seven-year-old had been raped and murdered. The police chief said the city was nearing anarchy and the state governor said that such disasters brought out the worst in people. Months later, the truth emerged. The "gunfire" had been a popping valve on a gas tank. Six people did die in the Superdome: four from natural causes, one from an overdose, and one by suicide. There was not a single rape or murder. The only looting was by groups searching for survival supplies. A fleet of boats came from distant cities to rescue people.

Another striking illustration of Bregman's optimism about humans is his example of the book *Lord of the Flies* by William Golding [16]. Briefly, the story is that a plane has gone down somewhere in the Pacific, and the only survivors are some British schoolboys who end up on a small, deserted island. At first, the boys are delighted to be on a storybook island and without grownups. They decide to live in a democracy, but this deteriorates as several refuse to carry out their tasks. A ship goes by, but their fire is out, and they cannot signal. They pick on each other, pinching, kicking, and biting. Weeks later, a British naval officer walks onto the island and finds three boys dead and the others with painted faces and no clothes.

Golding's book is a work of fiction and based firmly on the belief that self-interest is dominant in human behavior. Golding's book sold tens of millions of copies and was translated into thirty languages. He was even given a Nobel prize for his book.

Bregman decided that he would try to find a real case as close as possible to Golding's fictional story. It took some hard work and a little luck, but he found just the story he was looking for. Peter Warner was the youngest son of Arthur Warner, one of Australia's richest and most powerful men. Peter's father carefully groomed him to take over his father's business, but he ran away and went to sea to find adventure. He sailed the seven seas for a few years and returned home with a Swedish captain's certificate. His father was still not impressed and insisted that his son learn a practical profession. After five years of night school, he earned a degree in accounting and went to work for his father's business. The sea still beckoned to him, though, and he kept his fishing fleet in Tasmania. While on a fishing trip, by chance, he spied a small island: 'Ata. The island was once populated, but in 1863 a slave ship took all the inhabitants away. Since then, people believed the island to have been deserted.

Through his binoculars, Peter saw burned patches on the cliffs of the island. This was odd because fires rarely started spontaneously in the tropics. Peter steered his boat toward the island, and as he got closer, one of his men yelled, "Someone's calling!" Through his binoculars, Peter saw a naked boy with hair down to his shoulders leap from the cliffs into the water, followed by several others. Peter ordered his crew to load their guns, thinking the boys might be criminals dumped on the island. But as the first boy was pulled into the boat, he cried in perfect English, "My name is Fatai, and we reckon we've been on this island fifteen months." There were six of them, and a phone call back to Tonga verified that they were students at St. Andrews, an Anglican boarding school in Nuku'alofa, the capital of Tonga. At the beginning of their adventure, their ages ranged from 13 to 16.

How did these boys end up on a tiny, deserted island? Bored with their school assignments, they hungered for adventure and decided to escape to Fiji (500 miles away) or even more distant New Zealand. They had no boat, so they had to "borrow" one from a fisherman they disliked. Supplies to take on the trip were slim: two sacks of bananas, a few coconuts, and a small gas burner. They had no map or compass. None of the older boys knew how to sail, so they had to take 13-year-old Tevita, who had some sailing experience.

On their first day, the boys had smooth sailing with sunshine and a light breeze. But they made a big mistake that night by all falling asleep. They awakened to a thunderous storm with waves crashing over the rails. They tried to raise the sail, but the wind shredded it instantly. Then the rudder broke. For the next eight days, they drifted without food or water. Then on the eighth day, they spied possible salvation: a small island. There was no sandy beach or palm trees, only rocks rising more than a thousand feet above the water.

When Peter arrived, he found that the boys had "set up a commune with a food garden, hollowed-out tree trunks to store rainwater, a gymnasium with curious weights, a badminton court, chicken pens, and a permanent fire, all from handiwork, an old knife blade and much determination." The boys had agreed to work together in teams of two, with work divided into garden, kitchen, and guard duty. If there was a squabble, the opponents would retreat to the opposite ends of the island for four hours to cool, then come together to apologize. Their days began with a song and prayer. One boy made a makeshift guitar from driftwood, a coconut shell, and steel wire salvaged from their wrecked boat. Peter kept that guitar after the rescue.

The boys' stay on the island was no picnic. During the summer, there was little rain, and the boys suffered from thirst. They tried to build a raft to leave the island, but it fell apart in the surf crashing on the rocks. A storm dropped a tree, crushing their hut. One boy slipped off a cliff and broke his leg. The boys splinted it with sticks and leaves. While he was unable to work, the other boys did his work.

At the boys' return to Nuku'alofa, police met them and put them in jail because the stolen fishing boat owner was pressing charges. Fortunately, Peter came up with a plan to sell the boys' story to the media, although that plan never made significant income. The whole island welcomed the boys and Peter, though, with a big celebration, including a message from King Taufa'ahau. Peter later bought a new fishing boat and hired the six boys to see the world beyond Tonga. The boat's name was, of course, "Ata!"

The account of six boys from Tonga was the real-life version of Golding's fictional *Lord of the Flies*. It did not sell millions of books, and most people have never heard of the story, but it was a clear counter-argument to the widely held belief that people behave badly under challenging circumstances. The real boys became stronger working together under difficulties.

What makes people different in their relational behavior? On the one hand, there are kidney donors who risk major surgery and the loss of a backup kidney to improve or save someone's life, in some cases, not even a relative. At the opposite end of the scale are the psychopaths who seem to take pleasure in another human's suffering and death. The quality that measures this scale is empathy, the brain's intrinsic ability to experience how another person is feeling. Neuroscientists have determined that a part of the brain called the amygdala plays a major role in empathy and related decisions, along with a region in the frontal cortex. Studies showed that the super-altruistic kidney donors have an amygdala eight percent larger on average than the general population, while psychopathic criminals show reduced activity in the amygdala. [17,18] Empathy does not depend only on brain biology, however. Children who grow up in a stable and nurturing environment are more likely to demonstrate higher empathy. [19]

Trivers and others thinking about altruism considered a game studied in game theory concerning two individuals who might not cooperate, even if it appears that it is in their best interest to do so. The most well-known version of this game is the Prisoners' Dilemma. [20]

> "Two criminals are arrested and imprisoned. Each prisoner is in solitary confinement with no means of communicating with the other. The prosecutors lack sufficient evidence to convict the pair on the principal charge. They hope to get both sentenced to a year in prison on a lesser charge. Simultaneously, the prosecutors offer each prisoner a bargain. Each prisoner is given the opportunity either to: betray the other by testifying that the other committed the crime or to cooperate with the other by remaining silent. The offer is:
>
> - If A and B each betray the other, each of them serves two years in prison
> - If A betrays B, but B remains silent, A will be set free, and B will serve three years in prison (and vice versa)
> - If A and B both remain silent, both of them will only serve one year in prison (on the lesser charge)"

What should each prisoner do? The best solution to the Dilemma is, of course, for both prisoners to keep silent. Trivers later expressed the general best solution for dilemmas of this type as a recipe for reciprocal altruism in the form of a Golden Rule for Evolution: "Cooperate at the first step, then reciprocate thereafter." Another group expressed evolutionary altruism in the form of a question: "Is what I am doing benefiting me personally at the expense of the overall or critical functionality and/or operational wellbeing of the groups in which I am embedded?"

Mathematical biologist Martin A. Nowak and science writer Roger Highfield write in their book *SuperCooperators* about the importance of altruism in evolution. [21] There are many examples in biology of cooperation where a group achieves much more than single competitive individuals. Cells work together to create organs. Ants and other insects work together to achieve monumental feats of construction and food retrieval. But it is humans who have cooperated to achieve the most remarkable ends in construction, exploration, and technological development. Nowak and Highfield suggest that cooperation should be a third principle added to the evolutionary mechanisms of genetic modification and survival of the fittest.

Recently, researchers proposed an unusual possible source of altruistic influence: parasitic microbes. There are multiple known examples of these microbes

> ". . .making hosts perform reckless acts of irrational self-harm. There's *Toxoplasma gondii*, which drives mice to seek out cats eager to eat them, and the liver fluke *Dicrocoelium dendriticum*, which motivates ants to climb blades of grass, exposing them to cows and sheep hungry for a snack. There's *Spinochordodes tellinii*, the hairworm that compels crickets to drown themselves so the worm can access the water it needs to breed. The hosts' self-sacrifice gains them nothing but serves the parasites' hidden agenda, enabling them to complete their own life cycle." [22]

These negative examples led researchers to ask the question: Could parasitic microbes exert an altruistic influence? There is an example of microbes in a healthy human colon that drive intestinal cells to produce the neurotransmitter serotonin, circulating in the blood. Noone has yet carried out a biological test of this suggestion. However, researchers at Tel Aviv University have run computer simulations that analyzed interactions among members of a population over hundreds or thousands of generations. [22] The model in the simulations assumed that altruistic members incurred some fitness cost when they interacted, while the recipients of the altruistic acts benefited. The simulations included a mixture of virtual microbes that promoted altruism with others that did not. The researchers reported that "Over the generations, microbes that encouraged altruism in their hosts out-competed their rivals when both passed from one host to another and were subsequently passed from parent to child. This was true even when the population of 'pro-altruism' microbes was tiny at the outset." After many generations in the simulation, the host population mainly consisted of individuals carrying the altruism-promoting microbe.

SOCIAL CAPITAL

Sociologists have a term for the degree to which social networks are central to a society, and transactions are marked by reciprocity, trust, and cooperation: "social capital." The concept (not the term) was first used in the early 19th century by Alexis de Tocqueville as he described American life from a European perspective. He characterized Americans as prone to meet often to discuss issues of state, economics, or the world. He described this behavior as resulting from an emphasis on equality. Although Karl Marx used the term "social capital" in 1885, the term was first used in L. J. Hanifan's current meaning in a 1916 article [23] regarding local support for rural schools. Hanifan contrasted social capital with real estate, personal property, or cash.

In the 1990s, the concept of social capital gained popularity stimulated by a World Bank research project and several popular publications, such as the article "Bowling Alone: America's Declining Social Capital" and a book by a similar name, both written by Robert D. Putnam. [24] The article decried civil society's shrinking in the previous decades, at least partially due to people staying home and watching TV. Putnam's other possible causes for declining social capital were increased family mobility and women's entry into the workplace, reducing time for civic organizational involvement such as parent-teacher associations. Voter turnout had dropped by almost a quarter in the three decades before 1990. The number of Americans who had attended a public meeting on town or school affairs in the last year also had fallen by a third, with similar declines in those following a political rally or working for a political party. And "The proportion of Americans who reply that they 'trust the government in Washington' only 'some of the time' or 'almost never' [had] risen steadily from 30 percent in 1966 to 75 percent in 1992." People had also lost confidence in each other. The proportion of Americans saying that most people can be trusted dropped by more than a third between 1960 and 1993.

Why did Putnam give his article and book such an unusual title? The title came from his discovery that the number of bowlers increased by 10 percent between 1980 and 1993, while league bowling decreased by 40 percent. Bowling alone, indeed!

Social capital is, of course, critical to the functioning of a democracy – almost by definition. Besides, this form of capital is related to the two other forms of capital: economic and cultural. At first, it might seem that social and cultural capital would flourish in proportion to economic prosperity, workers having the time and resources to participate more in civic activities. Putnam reported, however, that the data indicated the opposite – that the economic attainments, quality of public life, and the performance of social institutions as well as education, urban poverty, unemployment, the control of crime and drug abuse, and even health all have better outcomes in civically engaged communities.

Putnam's writing stimulated much discussion about social capital for a decade or so after his publications but has declined since the mid-2000s. One reason has been the lack of an agreed-on definition for the term. In a 2004 blog, Tristan Claridge wrote, "The commonalities of most definitions of social capital are that they focus on social relations that have productive benefits. The variety of definitions identified in the literature stem from the highly context-specific nature of social capital and the complexity of its conceptualization and operationalization." [25]

Communications technologies have continued to expand rapidly since Putnam, now virtually eliminating geographical separation as a limitation to personal communications. It does not appear, however, that eliminating the geographical barrier has resulted in increased trust. In the past, individuals with a narrow interest or bias would likely have few supporters near them for reinforcement, resulting in a moderating effect on their hobbyhorse. Even if there are only a few people in their state, country, or even the world with a similar inclination, they could likely find each other for mutual support! The same communications tools also assist individuals with disruptive intent to spread "fake news" to increase distrust among individuals, groups, or companies. People with practical purpose can also connect more readily today too, of course, but trust is easier to tear down than to build.

HAPPINESS

Happiness is a universal human goal, and many paths have been explored to find it. What is known about the conditions most likely to result in human happiness? A remarkable study spanning nearly 80 years produced impressive answers to this question.

In 1938, researchers at Harvard began tracking the health of 268 Harvard sophomores hoping that a longitudinal study would reveal clues to healthy and happy lives. [26] The researchers later expanded their records to include the initial group's offspring, 1300 total by 2017, with most in their 50s and 60s. In the 1970s, researchers added 456 Boston inner-city residents to the study population, 40 still alive in 2017. Data recorded for all subjects included medical records, in-person interviews, and questionnaires. Also recorded were the broader events of their personal lives and careers, the valleys, and the mountaintops. As of 2017, 19 of the original group of sophomores were still alive, all in their mid-90s. All the subjects – the students, their children, and the inner-city residents – were known to have their basic needs met, so there were no hunger or homelessness issues, for example.

The researchers had a vast amount of data to sort through, but their general conclusions were clear:

> "The surprising finding is that our relationships and how happy we are in our relationships has a powerful influence on our health," said Robert Waldinger, director of the study, a psychiatrist at Massachusetts General Hospital, and a professor of psychiatry at Harvard Medical School. "Taking care of your body is important, but tending to your relationships is a form of self-care too. That, I think, is the revelation.
>
> Close relationships, more than money or fame, are what keep people happy throughout their lives, the study revealed. Those ties protect people from life's discontents, help to delay mental and physical decline, and are better predictors of long and happy lives than social class, IQ, or even genes. That finding proved true across the

> board among both the Harvard men and the inner-city participants." [26]

The results of this study were anticipated more than two millennia earlier in Plato's *Republic* as he describes his understanding of justice:

> "Life of just man is better and happier. There is always some specific virtue in everything, which enables it to work well. If it is deprived of that virtue, it works badly. The soul has specific functions to perform. When it performs its specific functions, it has specific excellence or virtue. If, it is deprived of its peculiar virtue, it cannot possibly do its work well. It is agreed that the virtue of the soul is justice. The soul which is more virtuous or in other words more just is also the happier soul. Therefore, a just man lives happy. A just soul, in other words, a just man, lives well; an unjust cannot." [27]

To sum up, humans can group in many ways to increase their quality of life. Regardless of the rationale that brings them together, however, true happiness depends on the participants choosing to make their links with others constructive. If even a few opt out of a group for personal advantage at the expense of others (Thrasymachus' justice), the group's efforts risk failure.

I have left one significant type of human organization out of this chapter's list: religion. That topic is the subject of the next chapter.

CHAPTER 18

Religion and the Supernatural

> Atheism turns out to be too simple. If the whole universe has no meaning, we should never have found out that it has no meaning... *C.S. Lewis.*

In all its varieties, religion is one of the oldest forms of human organization, following only behind the family. Anthropologists interpret cave paintings from the Upper Paleolithic era as appeals to spirits for success at hunting. The religious symbols moved from deep caves to pillars in sacred locations on the surface by the Neolithic period. Over the millennia, civilizations have put tremendous resources into land, buildings, personnel, and rituals for religion, indicating high importance for the activities. What can we say about the apparent human need for religion?

BENEFITS OF RELIGION

My late friend David Wilbur has written about this specific question in his book *Power and Illusion: Religion and Human Need*:

> "For millennia religions have offered us many wonderful images. These include the solitary worshiper; the moving public ritual; outspoken concern for the sick, poor, and disadvantaged; wonderful art and music; and magnificent buildings and grand centers of learning. These great religions would seem at first glance to be above criticism, but like all powerful human tools, they also have a dark side. Religion has been used to justify persecution, cruelty, terrorism, war, the suppression of knowledge, and

> extracting power and wealth from the struggles of the many. Such a wide range of uses cries out for a human explanation . . ." [1]

Wilbur lists the following areas where religion influences human lives:

Social Cohesion. Sharing common beliefs can be the glue that binds a group together, especially when those beliefs feature strong reinforcement for empathy and sensitivity to others' needs. As described in the previous chapter, there are many kinds of activities and interests shared to bring people together. Still, religion is unique in focusing each relationship on the other person, not inflating one's status or on objects or knowledge the other person has collected. In relationships with empathy, successes are shared and not envied. Failures elicit support and not superiority. Religion also serves to define group boundaries, distinguishing insiders from outsiders. Being inside the group boundary represents safety from the evils of the world.

Power Satisfaction. A strong leader can quickly become god-like within a community already predisposed to believing in a higher authority, becoming the visible definitive source of beliefs and rituals. Kings have sometimes claimed god status to reinforce their power and inspire their subjects. Throughout human history, religions have used hierarchical administrative structures to maintain authoritative power and strengthen a leader's control. A hierarchical class structure can then appear as sanctioned by God.

Existential Relief. Humans have a natural interest in and concern for the future. Most religions promise a supernatural afterlife where a believer can forget pain and fear. At the end of life, the believer is assured of continuing existence in a perfect environment. For the religious warrior, the future reward is the motivation for fighting with selfless abandon. The promise of a blissful future for the faithful is also a strong motivation for the believer's obedience to rituals in the present life.

Meaning Making. Religions teach the believer that the universe is not chaotic but purposefully created. The believer is placed on earth by God as part of the creation and has a vital role to play above other living creatures. Access to the supernatural is thus available to the believer. Like weddings, rituals -- part of family life -- serve as reminders of the universe's unique human status.

Personal Validation. A regular part of human growth and development is personal validation, first from parents and teachers and later from friends and careers. Adult confidence depends on a certain amount of reassurance from friends and coworkers. Being part of a religious community strongly increases the likelihood of this personal reinforcement.

Comfort in Distress. The world can sometimes seem like a scary place, especially with near-instant communication of every disaster and personal threat. Religion can assure that the deity is in charge and cares about the sufferings of earthly creatures. For example, a typical Christian reaction to illness or injury is to request divine intervention, for some, even in the form of a command. Excessive dependence on a supernatural deity can sometimes lead a believer to omit practical actions in a distressful situation.

Acculturation of the Young. Every parent wants their child to be happy, safe, and successful. When a family participates actively in a religious group, the traditional association and rituals can influence a child's development. Most religious groups organize some form of child education, whether as a supplement or a replacement for public schooling.

Identification of Cooperators. Human societies always have a few ready and even eager individuals to take advantage of other people for their benefit. A religious community with mores that discourage cheating is one way to help identify people and services that exemplify honesty in business relations. Some religious groups also encourage their members to use services offered by other members to strengthen their community's economic status.

Prayer, Meditation, and Problem Solving. Prayer is frequently a supplicating petition presented to a supernatural deity. Virtually all religions include prayer among their rituals, some with a highly stylized or prescribed form, while others leave the content up to the worshiper. Describing a problem or request in prayer can also benefit the petitioner and open the mind to a broader perspective. Meditation, often associated with prayer, can also be of value by clearing the mind and opening it to creative thoughts and problem solutions.

One of the most vigorous proponents of the importance of relationships in religion was the Jewish theologian Martin Buber [2], a man whose thinking appealed more often to Christians than his Jewish community. In an iconic treatise first published in German in 1923 *I and Thou* [3], Buber argued that what matters most is not an abstract understanding of God. Instead, the focus should be on entering into a relationship with God, accomplished by establishing genuine connections with other human beings, "I-Thou" relationships. By contrast, the relationships we have with people and things in the world for our benefit, Buber classified as "I-It" relationships. Buber did not denigrate this second type of relationship, but he argued that making these a dominant concentration would hinder a relationship with God.

Could religion be the form of human organization that steers all human interactions in positive directions? The Puritans arrived in the New World in the 1620s and 1630s with a vision to establish a society that would live by the Biblical instructions in Micah 6:8: "What doth the Lord require of thee, but to do justly, and to love mercy, and to walk humbly with thy God." Their future governor John Winthrop urged them to become an example to the world. "We shall be as a city upon a hill, the eyes of all people are upon us" (from Matthew 5:14). The New World colony did not turn out to be a model society. Still, vigorous evangelical preachers such as Jonathan Edwards (1703-1758) and Timothy Dwight (1752-1817) continued to pursue the goal of "shaping America."

Writing for *The Atlantic*, author Michael Gerson describes the ups and downs of evangelical influence in America:

> "In the mid-19th century, evangelicalism was the predominant religious tradition in America – a faith assured of its social position, confident in its divine calling, welcoming of progress, and hopeful about the future. Fifty years later, it was losing intellectual and social ground on every front. Twenty-five years beyond that, it had become a national joke." [4]

According to Gerson, several factors probably contributed to the deterioration of evangelical influence on American society in the latter half of the 19th century and the beginning of the 20th. There was mass immigration of Catholics and Jews in this period, diluting the percentage of Christian evangelicals in the population and the Civil War took a toll on social optimism. Intellectually, higher criticism of the Bible developed in this period, and evolution gained acceptance in science – both issues tending to decrease confidence in the prevalent literal interpretation of the Bible that was a pillar of evangelical belief. A final blow to evangelicalism in the mainstream was the Scopes "monkey trial" in 1925. In the trial, Christian politician William Jennings Bryan battled with Clarence Darrow against the right to teach evolution in a Tennessee school. Bryan formally won the case, but Darrow and evolution won the moral victory in the country.

In the mid-20th century, evangelicals fought back to claim the moral high ground by creating institutions – schools, outreach ministries, radio stations and broadcasts, and television broadcasts. Evangelists such as Billy Graham attracted large crowds and mixed with presidents and other politicians. Just as evangelicals were regaining moral influence, the sexual revolution began along with other major social changes, resulting in a widespread rejection of traditional beliefs. "As a result," continues Gerson,

> "[T]he primary evangelical political narrative is [presently] adversarial, an angry tale about the aggression of evangelical's cultural rivals. In a remarkably free country, many evangelicals view their rights as fragile, their institutions as threatened, and their dignity assailed. The single largest religious demographic in the United States – representing about half the Republican political coalition – sees itself as a besieged and disrespected minority. In this way, evangelicals have become simultaneously more engaged and more alienated."

The Puritans' high goal to establish a moral society through religion as an example to the world thus floundered on the shoals of human frailty.

I have described most of Wilbur's areas of religious influence on human lives as positive benefits. However, I acknowledge that religion can also be destructive. Education can become indoctrination and brainwashing. Service can become slavery. An influential leader can be a harsh dictator as well as a caring and inspiring person. Leaders can be divisive as well as cohesive. This book aims to emphasize the enormous creative potential of human groups when the relationships are supportive.

Many religions tend to focus on rituals and unique beliefs for many generations, forgetting the constructive ideas that inspired their creation. As a result, there has, in recent years, been a growing number of people who fall into a new category called "nones." When asked to report their religious affiliation, these individuals will select "none." They describe themselves as "spiritual, but not religious." This stance usually comes from cognitive dissonance and unresponsiveness of organized religions, but a desire remains for meaning and structure in their lives.

THE SUPERNATURAL

From the earliest records of humans' thinking about the world they live in, belief in the supernatural has existed – a universe not directly accessible by human senses. The earliest explanations for virtually everything that happened in the human world were gods, beings with human-like personalities but control over aspects of what we in the 21st century call the natural world. For these ancient people, a word like "supernatural" would not be needed because there was no need to distinguish between natural and supernatural. They lived in a world the gods controlled, and there was nothing more.

It was not until the beginning of experimental science and the Modern Enlightenment around the 17th and 18th centuries in the West that two conflicting sets of explanations began to vie for what occurred in the humans' physical world. Abrahamic religions maintained that God created the world and its contents while the early scientists began to describe laws that determined things like physical motion and electricity. Some early scientists, like Newton and Galileo, believed they were discovering how God worked. Others concluded that belief in a god or gods was no longer necessary once people knew the physical laws. Thus, the distinction between "natural" and "supernatural" was born.

In the 21st century, scientific knowledge of natural phenomena is vastly more significant than in the 18th century, but public belief in the supernatural remains. Some, especially atheists, echo a remark attributed to Laplace two centuries earlier: "I have no need of that hypothesis" (i.e., belief in a supernatural God controlling natural events). The public enjoys the benefits of new technologies resulting from discovering natural laws. But belief in the supernatural still exists. Examples are miracles, the world's creation, and a spiritual world paralleling the physical world (dualism).

As scientific understanding has developed, many – primarily religious people -- have tried to prove that the supernatural is still needed to explain some events. Miraculous healings attract people to religious meetings and shrines. Ronald Numbers in his book *The Creationists* [5] writes a comprehensive history of attempts to prove that God created the earth and life on the planet in supernatural events about six thousand years ago. Numbers begins his book by stating that "Within twenty years after the publication of Charles Darwin's *On the Origin of Species by Means of Natural Selection* in 1859, nearly every naturalist of repute in North America had embraced some [scientific] theory of organic evolution." Nevertheless, many Christians continued their belief in a recent and supernatural creation derived from a literal reading of the first chapters of Genesis. Some of these attempts to defend supernatural creationism used science's language and assumed the name Scientific Creationism.

One of the recent attempts to bolster belief in a supernatural creation was sparked in the mid-1990s by lawyer Phillip Johnson [6], becoming an Intelligent Design movement. Several individuals with impeccable scientific credentials wrote extensively in support of Intelligent Design. [7-10] They created a mathematical definition for design and, based on information theory, argued that the presence of design in nature necessitated direct supernatural intervention by God. The Intelligent Design promoters later deemphasized the religious origin of their argument. They presented it in strictly scientific terms (except for the assumption of supernatural intervention) so that public schools could teach it. Nevertheless, in the famous Kitzmiller v. Dover Area School District trial in 2005, after hearing presentations by scientists for and against Intelligent Design, Judge John E. Jones III ruled that Intelligent Design was not science and "cannot uncouple itself from its creationist, and thus religious, antecedents." Belief in Intelligent Design continues, however, supported by conservative Christians.

Acceptance of the supernatural is much more pervasive than just belief in miracles such as healings or protection from harm. Dominic Johnson, in his book *God Is Watching You* [11], describes several other common forms. The sports fan must wear a team hat or shirt to enable his team to score or the professional tennis player who will not step on the white lines. Tennis champion Rafa Nadal must have his hydration bottles lined up in a precise pattern, and he follows an elaborate sequence of hand motions before each serve. (Repeating the actual movements of a serve is entirely different, of course, as they represent learned muscle actions essential to a good serve.) United States presidents and candidates are not exempt from such superstitious rituals. Barack Obama had to play some basketball every morning during each of his election campaigns. Candidate John McCain carried a lucky feather and a lucky compass from the Vietnam war during his campaign, along with a lucky penny, nickel, and quarter. British comedian Shazia Mirza wore the same clothes at every gig for seven years to make sure the audience would laugh. And how many people will not walk under a ladder or expect something bad to happen on Friday the 13th. Each of these examples is a case of wishing a linkage between personal behavior and a physical event at a distance as much as thousands of miles away for the TV sports viewer. If asked, each of these individuals would probably admit that such a linkage is very unlikely. Nevertheless, rituals continue just on the chance that there might be an effect. (What do scientists really know, anyway?)

The word "supernatural" has different and distinctive meanings for the public and scientists. On the one hand, in the mind of the non-scientific public, a supernatural event is one that no one can duplicate on demand and for which experts have no explanation in terms of current knowledge. On the other hand, an event is supernatural for a scientist only if a public demonstration shows the event is contrary to natural laws, for example, breaking the law of conservation of energy. The problem is that scientists can only make such a demonstration within the limits of current technologies. Early, the only known sources of heat and light were hydrocarbons dug from the earth (coal, petroleum, peat, etc.) or burning dried plants. People would have treated light powered by a chemical battery or by mass to energy conversion as miraculous. It is essential when using the term supernatural to clarify whether the usage is in the popular or scientific sense.

There is a scientific and technological limit on categorizing any event as scientifically supernatural, and that limit changes with time as knowledge and technology improve. For this reason, all scientists operate under the principle of *methodological naturalism*. That means that when scientists search for an explanation of any observation or result of an experiment, they only postulate some combination of known natural laws or, in rare cases, a possible new natural law. It is never good science to insert the supernatural into an explanation for any observation, however rare or unusual. The supernatural's exclusion is not because it has been proven not to exist but because of a practical limit. Inserting the supernatural into any scientific explanation would automatically mean the arbitrary end of research in that area since the supernatural is the "universal solvent" that can "explain" anything!

By contrast, *Ontological naturalism* includes the belief that the supernatural does not and cannot exist. While there are people who espouse this idea, it is a philosophical belief with no support from or relationship to science.

RELIGION IN THE CATALOG OF THE UNIVERSE

In earlier chapters, I described how researchers expand understanding of the world by combining observation, experiment, hypotheses, models, and reduction in a cooperative community to minimize erroneous conclusions. What are the goals, origins, and future of religious beliefs? Do the roots of religions fit the steps I described for lower levels in the Catalog of the Universe? Religions have all the challenges described in the previous chapter for human sociology. Human complexity makes the results of religious groups challenging to characterize, much less predict. Ethical and privacy considerations impose severe limits on religion measurements, and experiment control is also minimal.

In the 1950s, a group came together at Harvard University to support scientific methods in studying religion. Started as a committee, the group developed into the Society for the Scientific Study of Religion. Here, from the Society's web site, is their description of their mission:

> "The Society for the Scientific Study of Religion (SSSR) is an interdisciplinary academic association that stimulates, promotes, and communicates social scientific research about religious institutions and experiences. ... SSSR fosters interdisciplinary dialogue and collaboration among more than a thousand scholars from sociology, religious studies, psychology, political science, economics, international studies, gender studies, and many other fields." [12]

At the University of Agder in Norway is a research project, begun in 2015, and unique in its effort to simulate religious trends. [13] The Modeling Religion Project, led by Wesley Wildman and LeRon Shults, is based on computer models that incorporate anthropological, archeological, psychological, and current demographic data related to religion. The project's goal is to conclude how and why religions have formed through history, their impact on individual and group behavior, and how they might develop in the future.

Some of the Modeling Religion Project data comes from surveys by religion and society researchers Chris Sibley and Joseph Bulbulia in New Zealand. [14] Aware of the human tendency to give more attention to religion in times of disaster, they surveyed New Zealanders' attitudes, values, and religious beliefs starting in 2009. There was another earthquake in New Zealand in 2011, so the researchers repeated their survey and compared the later results to those before the quake. "The findings showed that people living near the earthquake, whether religious or not before the event, became more religious in the wake of the tragedy, at least for a while."

CHOOSING A RELIGION

Most religions begin with a single individual, one who sees world problems and has ideas about solutions. These individuals' names are well known: Buddha, Abraham, Jesus, Muhammad, Zoroaster, and many other more recent founders of Christian denominations. (Hinduism and Shinto are examples of religious systems that claim no single founder.) After a religion has existed for a century or more, it can be challenging to know whether founders' declared characteristics are authentic or traditional products. Nevertheless, all religions with single or multiple identified founders claim supernatural inspiration for their founder(s). This claim assigns great significance to any statements from the founder(s) and establishes those statements as eternal truths, the religion's core beliefs.

All religions offer paths to achieving at least some of the benefits cataloged above. However, how these benefits are achieved can vary widely, as the techniques presented are always based on a worldview at least somewhat unique to the religion. Common themes in religious worldviews include the supernatural, one or more divine beings, creation of the world and universe by a divine being, and a continuation of existence after death. Religions distinguish themselves by their system of beliefs and their rituals.

What can a person in the 21st century seeking the benefits listed above make of the vast array of religious options vying for attention? Choose the largest group, the one nearby, the one practiced by parents, or attempt to evaluate the beliefs? If the latter, what standards could be used in the evaluation – since the religious systems themselves claim to set the standards! Claims of supernatural origin do not help make a choice since almost all religions make this claim, and there is no validation test.

A little lower than religion on the Catalog of the Universe, the field of medicine faced the same problem as recently as the 19th century. Many drugs and treatments offered high claims for success at treating ailments. We now know many of those offerings – like bleeding and potions containing arsenic and cocaine -- were not only incapable of curing disease but even harmful to health! But, at the time, there was no attempt to evaluate claims. A significant difference between medicine in the 19th century and today is that we have learned to ask for demonstrated results (even though some "patent medicines" and unproven treatments still attract desperate sufferers). Medicine has moved beyond tradition and random trial and error for discovering remedies. The emphasis now is on mechanisms in the body -- how the body systems work in health, what goes wrong in disease, and how to eliminate disease origins.

Could a similar approach, looking at mechanisms and results, be used to evaluate or even improve the benefits of religion? There are obvious problems, both logical and practical. In the case of health, it is not hard to get an agreement on what constitutes good health. In religion, the systems themselves usually define what is "good," and in some cases, the definition might include uncertainty or physical danger and suffering. While the body systems are complex and far from completely known, human behavior is even more variable, complicated, and challenging to evaluate. "Miracles" are of dubious worth as evidence of the supernatural (in the scientific sense) because, as I have argued earlier, the limits of what can be done within the "natural" change continually.

In practice, people usually choose their religion for reasons that are not strictly logical or even conscious. Family and environment while growing up have strong influences, sometimes positive and sometimes harmful. Scientific observations can be shared and experiments repeated by others for validation. There is no objective way to evaluate the validity of the supernatural elements of any religion. A person's feelings and experience are typically the deciding factors, especially when there is an ecstatic experience.

In the previous chapter, I described how certain contemporary philosophers argue that it is unnecessary to posit the supernatural as the source of moral principles for society. Atheistic humanism can be a good source through rational thinking. I believe there is a practical flaw in this argument. If a community was composed entirely of educated and thoughtful individuals willing to live by the best-known principles, the atheist idea might have a chance to succeed. However, real societies are never uniform in population, much less uniformly populated by thoughtful and consistent philosophers! Such individuals form one tail of a distribution with an opposite tail of individuals who care little for social mores, with most of the population somewhere in between. For the self-centered group and many in the middle distribution, belief in a supernatural God defining right, wrong, and consequences of choices is probably essential to provide stability in a society.

In his book *Why We Need Religion*, author Stephen Asma considers the emotional, rather than the strictly rational, aspect of human life. [15] Each day, we face many choices and situations to react to. These can be as simple as greeting a family member in the morning or as complex as facing a challenging personnel problem at work. Our first response to most of these situations is automatic, taking no time to go through a logical process of gathering and weighing the evidence for and against each option. These automatic emotional responses develop unconsciously and gradually through life based on our experiences, choices, and consequences. Emotional responses may aid or worsen their triggering situation. Asma discusses different emotional problems common to humanity, for example, sorrow, anger, fear, forgiveness, peace, joy, and play. Do we have to live with our automatic reactions to these situations, or can they be modified? Asma's position is that humans can alter and exert some control over their emotional states by choice and effort. All religions have developed methods for modifying and controlling emotional responses: meditation and prayer to quiet mental turmoil, morals to guide choices, beliefs to establish a worldview including an afterlife to allay existential fear. The efficacy of a religion's methods does not necessarily depend on the objective truth of a religion's beliefs. Atheists who ridicule religion as irrational and unnecessary ignore the central role of human emotions.

RELIGION AND THE SCIENCES

Based on what we can learn from the Catalog of the Universe, the main subject of this book, and the desire to make religion more positive, I would like to offer three suggestions.

First, any religion worth considering should, I believe, have as its goal human flourishing. Whole books have been written on the meaning of "human flourishing," [16] so my definition will have to be very short and incomplete: human flourishing means maximum good health, personal freedom within a society, creativity, and opportunity for development for all humans within an imperfect environment. In terms of the Catalog of the Universe, this result for human groups will require positive and reinforcing human relationships. It is no coincidence, I suggest, that most religions contain within their code of beliefs some version of a "golden rule": treat others as you would be treated. Even evolution now includes a golden rule.

Second, I suggest that it is valid to look at the results produced by a religion to see how well that system of beliefs supports and improves human flourishing. Any such evaluation will be subjective, of course, but with the vast amounts of data now being collected, there are opportunities for specific research. Any system that separates a part of the human population for different and lower standards of flourishing or ignores the human effect on the environment would receive a more inadequate evaluation by my criteria.

Third, I suggest that religions could learn from the sciences farther down the Catalog and use the reduction method to sharpen their own beliefs for better results. Specifically, that would mean looking at psychologists and sociologists' research to learn more about improving human relationships and reinforcing positive relationships. Religions might even conduct sociological or psychological experiments of their own for the same purpose. Religions have tended to depend mainly on tradition and subjective experiences, but traditional methods can quickly become outdated and emotional experiences are very personal in a rapidly changing world.

Religion and the sciences began in a close relationship. The founders of modern science, such as Kepler, Boyle, Newton, and Bacon, were religious men and considered their natural world studies as learning about God. Religion made contributions to the origins of scientific thinking, such as the idea that God exerted a causal influence on the world through natural laws. As scientific studies expanded in the 19th century, differences began to arise, for example, from Darwin's famous publication regarding human origins. Example publications pointing out the split are John William Draper's *History of the Conflict between Religion and Science* (1874) and Andrew Dickson White's *History of the Warfare of Science with Theology in Christendom* (1896). Examples cases cited by Draper include Galileo's conflict with the Catholic Church, Hypatia's death at the hands of a Christian mob, medieval belief in a flat earth, papal ex-communication of a comet, the Church's ban on dissection, Copernicus' dethroning of humanity (by removing humans as the center of the universe), and Bruno's execution as a martyr to science. As described in a recent book edited by Ronald L. Numbers *Galileo Goes to Jail and Other Myths about Science and Religion*, some of these events were mythical. [17] The conflict intensified in the early 20th century with *The Fundamentals: A Testimony to The Truth* (usually known as *The Fundamentals*), 90 essays printed in 12 volumes between 1910 to 1915 by the Bible Institute of Los Angeles. The purpose of the series was to affirm conservative Protestant beliefs, including a recent fiat creation by God.

The religion-science conflict has continued in episodes since the early 20th century, a concrete example being the efforts to teach a recent fiat creation and biological evolution in public schools, actions that continue as this book is being written. Much of the interaction between the two sides occurs in one-sided polemics. An example was the debate between science educator Bill Nye defending evolution and Christian fundamentalist Ken Ham supporting a recent fiat creation. [18] Paleontologist and historian of science Stephen Jay Gould in a 1997 essay suggested a compromise by proposing the two sides to be "non-overlapping magisteria" (NOMA). [19] He described the NOMA principle as "Science tries to document the factual character of the natural world and to develop theories that coordinate and explain these facts. Religion, on the other hand, operates in the equally important, but utterly different, realm of human purposes, meanings, and values—subjects that the factual domain of science might illuminate, but can never resolve."

The Catalog of the Universe proposed in this book illuminates the religion-science relationship, first by showing that both exist in a hierarchy of levels of human understanding of the natural universe. The term "science" typically refers to the lower, middle, and cosmic levels in this Catalog, representing the study of physical aspects of the universe. "Religion" claims a narrow but essential slice high in the Catalog, along with other human social groups. Both science and religion study areas in which relationships determine outcomes. Methods of study in science and religion differ because the observations and experiments are more repeatable and exact in science. Observations are much more challenging, variable, and personal in religion. Understanding in scientific fields has benefitted from the reduction method, but religion rarely has utilized that method of study. The scientific fields dealing with the highest levels of complexity, e.g., medicine and sociology, have been slow to develop quantitative theories, depending instead on qualitative conclusions and hunches. Similarly, religion's theory equivalents – sets of doctrines and beliefs – are usually derived from authority and are generally not evaluated for their results.

The sciences and religions are both components of the Catalog of the Universe, so there should be no reason for animosity between them. Gould's NOMA concept is partly true because religion studies a different area from ecology and physiology. The methodologies of religion are different from the biochemistry methods because the two are vastly different in complexity. Religion could benefit from some of the study methods of science, such as experiment and reduction. On the other hand, scientists could be more respectful of the challenges religion faces in understanding and modifying very emotionally charged areas of human behavior.

Two scientific authors studying complexity in nature have used the word "sacred" to describe their reactions to what they observe. Stuart Kauffman, a leading researcher in biocomplexity at the Santa Fe Institute in the 1990s and the founding Director of the Institute of Biocomplexity and Information at the University of Calgary, entitled one of his books on emergence and complexity *Reinventing the Sacred: A New View of Science, Reason and Religion.* [20] In this book, he describes his profound respect and awe for the amazing creativity and complexity he sees in the universe, from molecules and cells to the vast structures of the cosmos and human cultures – as I have described in this book as a Catalog of the Universe. Kauffman is not a believer in the traditional God of Christianity. However, he suggests:

> "It may be wise to use the word *God*, knowing the dangers, to choose this ancient symbol of reverence and anneal to it a new, natural meaning. God is our name for the creativity in nature. . . Using the word *God* to mean the creativity in nature can help bring us the awe and reverence that creativity deserves. . . This sense of God enlarges Western humanism for those who do not believe in a Creator God. It invites those who hold to a supernatural Creator God to sustain that faith, but to allow the creativity in the universe to be a further source of meaning and membership. I hope this sense of God and the sacred can be a safe, spiritual space we can all share." [21]

In his book *The Constant Fire: Beyond the Science vs. Religion Debate* [22], astrophysicist Adam Frank also builds on the word *sacred*:

> "Does what we call sacred have any connections with what happens in science? I believe the answer is yes, as long as we do not turn the experience of life's sacred character into the *sacred* as a substitute for *divinity*, and as long as the emphasis remains on what we encounter in experience. Can we unpack "the sacred" and use it to frame a new view of science and religion?" [23]

Frank's answer to this question is "yes," and in his book, he tells the story of developing human worldviews from the ancient gods to current science. Through the many centuries, humans have experienced the fire of wanting and needing to understand the universe they inhabit and their place in it. Frank concludes:

> "If we locate science alongside the field of human spiritual aspiration it becomes an instrument for gaining both knowledge and wisdom. This is the fundamental change. When science and myth are drawn into a parallel complementarity, we can see both as a way to affirm life and its place in the cosmos." [23]

(Myth in this last sentence does not mean "fairy tale," but rather the story and history of a people, the origin of their beliefs.)

Thus, Kauffman and Frank challenge religious believers – not to replace their religious beliefs with science, but rather to embrace them both for a richer and more comprehensive understanding.

Science exists to benefit humans, to make their lives safer and more pleasant. Nevertheless, some humans reject the conclusions of science in favor of theories of questionable origin. The next chapter discusses this aberration.

CHAPTER 19

Why There is Distrust of Science

> "Misperceptions...have become almost as difficult to overcome as diseases themselves." *Reed V. Tuckson* [1]

Science is the process of observing the natural world to discover patterns and principles that can benefit humans. If this is so, why does distrust in science exist?

Measles is a highly contagious viral disease that causes flu-like symptoms and a rash. Complications occur in 3 out of every 10 cases; young children are especially vulnerable. [2] Before the measles vaccine was available, nearly every child got measles by age 15. Each year, 3 to 4 million people contracted measles resulting in about 500 deaths and 48,000 hospitalizations. The measles vaccine became available in 1963. By 2000, immunization almost eliminated measles in the US. But then, measles cases began to rise again, and in 2019, the US experienced the most significant number of measles cases since 2000.

Why did the yearly case numbers rise again? A theory spread between 2000 and 2019 that the measles vaccine was dangerous, risking autism and death. As a result, mothers began to refuse vaccinations for their children. The theory originated from a paper published in The Lancet in 1998 by Andrew Wakefield and colleagues. [3] They conducted a clinical study using the MMR vaccine (measles, mumps, and Rubella) and concluded that the vaccine increased the risk of autism in their test group. The press latched on to this result and published it widely, causing many mothers to refuse the vaccine, even when school children were required to be vaccinated. The result was that measles cases began to rise again.

A repeat of the study with more children (Wakefield followed only 12 children) refuted Wakefield's conclusion. Ten of Wakefield's twelve co-authors on the original paper also wrote later that their initial conclusion was wrong. The Lancet completely retracted the publication in 2010, citing defects in the experiment design. Unfortunately, a bell had been rung, and it could not be un-rung. The fear of vaccines did not wane but has continued to grow with the support of conspiracy theorists and social media.

Other similar false theories about contagious diseases have also appeared. An estimated 350,000 additional HIV deaths occurred in South Africa because of neglected treatments caused by the country's elected president's conspiracy theories. I am writing this book during the COVID-19 virus pandemic of 2020. At this time, more than 40% of the US population say in surveys that they will not take a COVID-19 vaccine when it is available.

This book's message is that science, a combination of observation and theory, is the best-known method of understanding humans' environment and applying that understanding to benefit people. Science is a human activity and thus subject to occasional mistakes and even willful deception. Researchers make up a self-checking community with repetitions of experiments and reviews before publication. A consensus is always the goal for any issue. That being the case, why would people reject science for alternative ideas that have no rational basis?

Contempt for science is not new or limited to a particular economic or political group. [4] C.P. Snow, in his famous 1959 lecture on "The Two Cultures and the Scientific Revolution," commented on the disdain for science among educated Britons and called for greater integration of science into intellectual life. Philosopher Thomas Kuhn wrote that science busies itself with solving puzzles before lurching to some new paradigm that renders its previous theories obsolete; indeed, unintelligible.

Why is it that a significant percentage of the public is ready to accept a conspiracy theory over scientific consensus views on such topics as vaccines, evolution, or global warming? Here are some reasons why the public persists in anti-science positions and misconceptions about the nature of science that distort the public opinion of a carefully researched scientific consensus.

Tribal loyalty among humans is a powerful force. Communities can become echo chambers, amplifying both bad ideas and untrustworthy data. The problem is worse because we often care more about conforming to our group's norms and narratives than discovering the truth. Filippo Menczer and Thomas Hills have studied how algorithms and manipulators exploit our cognitive vulnerabilities: [5]

> "Social groups create a pressure toward conformity so powerful that it can overcome individual preferences, and by amplifying random early differences, it can cause segregated groups to diverge to extremes. . . Social media amplifies homophily [following others who are like us] by allowing users to alter their social network structures through following, unfriending, and so on. The result is that people become segregated into large, dense and increasingly misinformed communities commonly described as echo chambers."

Menczer and Hills also showed that bots – automated accounts that impersonate human users – significantly reduce the information quality in a social network.

A recent study looked at the intensity of hate toward the US's opposite political party over the past 40 years. [6] The feeling toward the opposing party was nearly neutral in 1980, but the hatred has increased steadily until the authors published their paper in 2020. Other-party hate has emerged as a more potent force than in-party love. If a tribal leader expresses an anti-science position on any issue, members of the tribe (political party, for example) will be very reluctant to deviate from the leader's views as the deviation might well cost them their membership in the tribe.

Humans also have a **fascination for positions contrary** to society's assumptions. For example, if a YouTube video questions science, it gets substantially more views than explanations of science. [7] A Brown University professor found that over three decades, opposition to climate action was twice as likely to get news coverage as support for climate action. [8] Entities have also taken advantage of this fascination. According to Naomi Oreskes and Erik M. Conway, in the first half of the 20th century, most Americans believed that science made their lives better. The tobacco industry then started promoting the idea that the link between smoking and cancer and other diseases was uncertain or incomplete. The attempt to regulate tobacco was a threat to American freedom. The industry made products more addictive by increasing the nicotine content while denying that nicotine was addictive. Proponents for delaying action on acid rain, the ozone layer, and climate change also used the same argument. [9] Now, the target for this spurious "freedom" argument is mask wearing. [10]

Another contributing factor is that **the public sees science as a search for absolute scientific "truth."** If "truth" means an objective statement of the nature of the physical universe, there are very few "truths" in science. The best that scientists can do is record observations and then build an approximate understanding of the natural world based on those observations. Even with remarkable new instruments and computers, recorded data are still a limited snapshot of reality. There is no way to prove that additional information would not change the current view. Basing a theory of nature on experiments and observations does maintain a connection with reality. As part of the scientific method, scientists have practices designed to minimize observation and theory limitations. For example, repetition of an experiment is always necessary, not only by the initial observer but also by others in the scientific community.

Any new scientific theory is not complete until the scientific community debates it, reaching a **consensus**. The process includes replicating experiments under an appropriately wide range of conditions, and there are detailed comparisons of the new theory with previous ideas and related concepts. The more diverse the participants in the discussion and the broader the range of disciplines brought to bear, the greater the weight given to the consensus. A conclusion reached in this way is not easy to overturn, and rightfully so.

The public can easily have quite a different view of the consensus process. A consensus is a majority view, but the scientific community's 100% agreement is not always possible. The public may still reject the scientific community's overwhelming conclusion without understanding uncertainty in science or because of a critical argument from one or a few detractors. The public will not give correct weights to the community consensus and the detractor's view and conclude that the scientific consensus is not valid. The media may also reinforce the public's mistaken conclusion by erroneously giving equal weight to dissenting opinions — all in the media's attempt to present a "balanced" view.

Some may reject a consensus because they suspect social, political, or religious motives. The public expects advocates of all kinds to present selected or even manufactured "truths," and they can easily regard the scientific consensus in the same way. Uncomfortable scientific conclusions are dismissed as attempts to advance a hidden agenda. An example is the materialism and atheism charges leveled at evolutionary biologists by those who see evolution in conflict with the Bible.

Public detractors of science often raise the issue of **uncertainty in science**. There are multiple sources of uncertainty in the scientific process: measurement precision, too little data, inherent randomness in the process under study, and an incomplete understanding of the natural mechanisms involved. Uncertainty is thus always present in attempting to predict natural events or outcomes. Scientists estimate and reduce errors by various means, such as statistical tests, improving instrumentation, and repeating experiments. A trustworthy scientific report describes both the conclusions from the observations and the uncertainty in those conclusions.

The public can easily be made uncomfortable with science by quoting uncertainties at decision time. A scientific theory that is not "proven" cannot be trusted is the assumption. But scientific theories of natural processes are never proven in an absolute sense. If someone claims to know something for sure, the only thing certain is they are trying to fool you! A theory's value grows with repeated successful testing. Waiting for science to achieve an unobtainable level of certainty only results in a delay. Even an idea with some uncertainty is a better basis for decision-making than rumors, unsubstantiated claims, or outright guessing. Failure to act is also a decision that may have consequences.

Scientific theories may seem by lay persons as indistinguishable from science fiction, guesses, or speculations. Scientists always base their theories on observations, however. Imagination and creativity are required to find the patterns, but data must constrain conclusions. Also, theory building is not considered complete in a single attempt. A good theory is tested repeatedly with new experiments and observations. Some approaches may fail, but each successful test brings increasing confidence in a surviving theory.

Any process **will act only within a particular range of time scales.** Trends can only be recognized and understood in the context of a specific time and space scale. This principle is critically applicable to the current hot topic of climate change. No single climate event or even an unusual season can by itself be considered evidence of climate change. For example, in North America, the Winter of 2009-2010 was frigid. However, by itself, that observation is not evidence for or against global climate change because it refers to a limited region. Only trends studied for years or longer can be valid evidence for or against climate change.

There are at least six predominant mechanisms that determine global climate: solar radiation, plate tectonics, ocean circulation, atmospheric composition, albedo, and human causes. These mechanisms have effects at time scales ranging from years to billions of years. Variations from some can produce effects on widely different time scales. For example, ocean circulation can be at the root of changes on a scale of years (El Nino and La Nina oscillations) up to tens or hundreds of millions of years (changes in ocean basin's shape or connections).

Scientists are human, so it should not be surprising that **some misbehavior by scientists occurs,** some willfully and some by neglect or ignorance. Scientists have strong motivations to succeed, answering an important scientific question, for example, and publishing regularly. Most also need to raise money to support their work. A few are willing to cut corners to achieve these goals. Stuart Ritchie has written a book titled *Science Fictions: How Fraud, Bias, Negligence, and Hype Undermine the Search for Truth.* [11] in which he describes examples of each of these misbehaviors and how scientists should protect against them.

The US Department of Health and Human Services (HHS), responsible for all public funding of research related to human health, operates the Office of Research Integrity (ORI). This office sets ethical standards for all research funded by HHS. ORI requires each research site to have a local policy and monitor research integrity and report yearly to ORI on any cases investigated.

Educators, public officials, and scientists can do things to **increase the public's trust in science.** [12] The most obvious step is to increase scientific literacy and clarify the goals and methods of science beginning in education. Support for science by public officials at all levels of government can be a positive influence. The opposite is true, as well. Trust in medical scientists is lower among blacks than in Hispanic and white adults, so addressing racial issues can also increase trust. Making scientific information public assures that there are no hidden agendas, as can advocation for the role of independent science in decision making. Scientists can join advocacy groups, testify at hearings, and communicate with local and state leaders. Support for science must be active and not limited to defending against overt attacks.

With the measles vaccine example in recent history, anyone wondering if they should follow scientific recommendations have enough evidence to make the rational choice. But the likelihood is that the current generation will forget the lessons learned earlier. Many will succumb to rumors without foundation and miss the hard-won benefits from scientific research.

CHAPTER 20

Science Enriches Human Life

> How intimately the social and religious emotions are connected with this primary fact of the mutual dependence of two human beings, and how from it slowly emerge all the marvels of Art and Science. *George Henry Lewes (1874).*

Early in the development of science, even nonscientists realized that this activity could be of benefit to people everywhere. In 1782 Thomas Paine wrote:

> "Science, the partisan of no country, but the beneficent patroness of all, has liberally opened a temple where all may meet. Her influence on the mind, like the sun on the chilled earth, has long been preparing it for higher cultivation and further improvement.
>
> The philosopher of one country sees not an enemy in the philosopher of another: he takes his seat in the temple of science and asks not who sits beside him." [1]

It has been necessary for humans throughout their existence to understand the environment they live in. Multiple motivations drive this desire: needs for food and shelter from the elements, safety from animal and human predators, association with other humans – friends and family, and certainty to face an unknown future. Western science arose late in human history and replaced practices based solely on religious beliefs and philosophy. Establishing life practices and world theories on observations and experiments now seems obvious, but the change required a new way of thinking.

The new understanding generated by science has produced significant changes in the way humans live. For the first three centuries, the focus was on collecting observations – natural or from experiments. By the 19th century, theories were receiving attention, and much of the focus was on energy in different forms – mechanical, thermal, electrical, chemical – and conversion between them. This understanding led to machines that reduced human physical labor and increased travel both for humans and goods. Later, new knowledge about the human body changed medical treatments from bleeding and potions based on cocaine and arsenic to drugs and procedures that reduced human suffering and directly attacked illness causes. Electronics began in the early 20th century, initiating new worlds of communications and information storage and retrieval. These developments changed human lives and societies, sometimes thoughtfully and for human betterment, but sometimes selfishly for personal gain resulting in harm to many.

My objective in writing this book has been to utilize complexity to show how all the human's fields of study, including themselves and their surroundings, fit into a pattern. In a single hierarchy, all the subjects usually called sciences are levels, including ecologies and human social activities from education and government to the arts and religion. I think E. O. Wilson had a comprehensive view like this in mind when he wrote:

> "Every college student should be able to answer the following question: What is the relation between science and the humanities, and how it is important for human welfare? Every public intellectual and political leader should be able to answer that question as well." [2]

The Catalog of the Universe described in this book clarifies common themes and methods in humans' exploration of themselves and the universe they live in. From this inclusive organization of activity and knowledge, some general conclusions were possible in earlier chapters.

Complexity as an Organizing Principle: We have seen that complexity is the basis for a hierarchy that links all types of objects and groupings in the known universe, from elementary particles to galaxy clusters – from quarks to the cosmos. The Catalog includes human groups from family to business, education, government, arts, and religion. It also clarifies the similarities and differences between studies at all the many levels of the hierarchy.

Relationships are Key: All levels of the Catalog of the Universe depend on relationships for their existence and unique qualities. To borrow a 19th-century phrase, it is "relationships all the way down!" The success of science depends on this fact to build theories. Human organizations of family, government, business, and education function better when everyone strives to make their relationships positive. Understanding all of science as a hierarchy helps us to appreciate the benefit of multidisciplinary research teams. Reduction as a research method is more powerful, and creativity is more fruitful when knowledge of neighboring Catalog levels exists within a research team. Diverse research groups have become more common since the middle of the 20th century.

Synthesis and Creativity: Western science began in the 16th century and advanced slowly as late as the early 20th century because of the isolation of research fields and the theory limitations. But scientists have been creative and continue to develop new methods, for example, moving to multidisciplinary teams and computers for theories as I have described in this book and creating many new measurement techniques. Synthesis is beginning to receive more attention because it is synonymous with creativity, not only in science and technology but also in the arts and other human activities. 20th-century philosopher Paul Feyerabend reminds us, however, that scientists must not rely solely on advancing current paradigms [3]. Advances in classical physics would never have led to quantum mechanics, for example. However distasteful it might be to tolerate radical ideas, they must be allowed to remain on the table.

Science and Religion: The Catalog of the Universe described in this book includes every branch of science and all religions and other human organizations in one unified structure. This structure makes it possible to compare and contrast the sciences and any religion on appropriate terms. Religions are organizations of humans; their ability to provide benefits to humans depends on their members' quality of relationships, just as in the rest of the Catalog. If ever, religions have rarely taken advantage of the method of reduction: studying the human connections the religion's belief system produces as sociologists do to find ways to further the benefits desired for their adherents. On the other hand, scientists should respect the challenges of human complexity faced by religions in forming and teaching morals and providing hope for the future.

There are aspects of human thinking and concern that science cannot verify or even address. While the Big Bang theory may describe the origin and development of the universe we live in, the idea itself assumes the existence of natural laws as we observe them now. Where did these laws come from? Is there any continuation of life after death? Is our universe the only one, or do other universes exist? These questions are outside the reach of science because science is limited to understanding the physical and temporal reality humans live in. Religion and philosophy may address the big questions. Still, the wide range of answers offered indicates that the only "proofs" possible in these areas originate from human experience and emotions.

Have humans benefitted from the vast increase in knowledge and understanding of the natural world? There are, without question, vastly more opportunities for human comfort and creativity because of scientific understanding and the resulting technologies. Unfortunately, a significant fraction of the population continues to cling to beliefs from the era of the Great Chain of Being, for example, a static, unchanging world and universe. A significant population is rejecting even the persuasive powers of reason and evidence. Contemporary philosopher Rebecca Goldstein writes

> "Tribalism and authoritarianism are reasserting themselves across the globe. These, too, are means for trying to gain our bearings, which, being primitive, come to us far more naturally and forcefully than do scientific and philosophical reason. Tribalism and authoritarianism are where our species began, and, over the course of our long history, almost every variation of them has been tried. The outcomes, even when not disastrous, are never as conducive to human flourishing as the states of affairs to which reason has brought us. With so much to lose, we need to marshal the full resources of human reason." [4]

It is now becoming abundantly clear that humans can use the new powers of science and technology for creative or destructive purposes, the latter damaging humans and other forms of life and even the earth itself. Ultimately, the uses of science are determined by human relationships of all types, from family, education, government, and commerce to the arts and religion. Unfortunately, all these human systems depend primarily on tradition and thus are slow to change, falling behind in profitably utilizing new opportunities. Hopefully, recognizing that constructive and trust-encouraging relationships are the key to flourishing can motivate all humans to impact every relationship positively.

I end this book with a quote from Nikola Tesla, the electrical genius and creative inventor:

> "Though free to think and act, we are held together, like the stars in the firmament, with ties inseparable. These ties cannot be seen, but we can feel them . . .
>
> For ages this idea has been proclaimed in the consummately wise teachings of religion, probably not alone as a means of ensuring peace and harmony among men, but as a deeply founded truth. The Buddhist expresses it in one way, the Christian in another, but both say the same: We are all one . . . Science, too, recognizes this connectedness of separate individuals, though not quite in the same sense as it admits that the suns, planets, and moons of a constellation are one body, and there can be no doubt that it will be experimentally confirmed in times to come, when our means and methods for investigating psychical and other states and phenomena shall have been brought to great perfection." [5]

REFERENCES AND NOTES

Prologue

1 After I graduated from college, the requirement for calculus was first moved from the sophomore year of college to the freshman year. Still later, beginning calculus was moved down to the high school level.

2 Mention "vacuum tubes" to someone of the current generation and you are likely to get a blank stare in response. A "vacuum tube" was a small, sealed class cylinder enclosing a vacuum and two or more metal electrodes with external connections. From the early 20th century when they were invented, through the 1950s, tubes were the only non-mechanical devices available for generating, amplifying, or switching electronic signals. In the 1960s, solid state transistors took over most of the functions of vacuum tubes because they used less power, were much smaller, and were cheaper to make. By the middle of the 1960s, manufacturers started making integrated circuits: multiple transistors in a circuit on a single semiconductor chip. By 2020, integrated circuits are made with billions of transistors and have become foundations for smartphones, tablets, laptops, and computers of all sizes.

Chapter 1

1 Mark, Joshua J. (2018, March 23). *Ancient History Encyclopedia: Religion in the Ancient World.* Retrieved from http://www.ancient.eu/religion/.

2 Lovejoy, Arthur O. (2009) *The Great Chain of Being: A Study of the History of an Idea.* New Brunswick: Transaction Publishers. A reprint of the original 1936 publication with a new Introduction by Peter J. Stanlis.

Chapter 2

1 Goldstein, Rebecca N. (2017, December 11). The Scientists and the Philosophers Should Be Friends. *Free Inquiry,* 38 (1).

2 Laplace, Pierre Simon (1902). *A philosophical essay on probabilities,* Translated from the Sixth French Edition. 4. New York: John Wiley & Sons.

3 "History of the Sears Catalog." Sears Archives. Retrieved from http://www.searsarchives.com/catalogs/history.htm.htm .

4 "Sears." Wikipedia article. Retrieved from https://en.wikipedia.org/wiki/Sears .

Chapter 3

1 Cockell, Charles S. (2018) *The Equations of Life: How Physics Shapes Evolution.* 83. New York: Basic Books.
2 McGowan, Kat (2013). What Makes You So Special. *Nautilus,* 1. Retrieved from http://nautil.us/issue/1/what-makes-you-so-special/cooperation-is-what-makes-us-human.
3 Gefter, Amanda (2015). Quantum Mechanics is Putting Human Identity on Trial. *Nautilus,* 30. Retrieved from http://nautil.us/issue/30/identity/quantum-mechanics-is-putting-human-identity-on-trial .

Chapter 4

1 Haeckel, E. H. P. A. (1866). *Generelle Morphologie der Organismen.* Berlin: G. Reimer.
2 Mayr, Ernst. (1982) *The Growth of Biological Thought: Diversity, Evolution, and Inheritance.* 64-66. Cambridge, Massachusetts: The Belknap Press of Harvard University Press.
3 Wikipedia: "Works of Francis Bacon", n.d.
4 Comte, Auguste. (1908) *A General View of Positivism.* Translated from the French by J. H. Bridges; a new edition with an Introduction by Frederic Harrison. Kindle Location 771-774. London: George Routledge & Sons Limited. EBook version provided by Project Gutenberg, 2016.
5 Salthe, S. N. (1985). *Evolving Hierarchical Systems: Their Structure and Representation.* 175. New York: Columbia University Press.
6 Needham, J. (1932) (Quoted in Reference 2)
7 Woodger, J.H. (1929) *Biological Principles: A Critical Study.* 449. London: Kegan Paul.
8 Koestler, A. (1967) *The Ghost in the Machine.* 45. London: Arkana.
9 Burger, William C. (2016) *Complexity: The Evolution of Earth's Biodiversity and the Future of Humanity.* Amherst, New York: Prometheus Books.
10 Simon, H. A. (1962) The Architecture of Complexity. *Proc. Am. Phil. Soc.* 106 (6),467-482. Retrieved from www.jstor.org/stable/985254.
11 Simon, H. A. (1969) *The Architecture of Complexity: The Sciences of the Artificial.* Cambridge, Massachusetts: MIT Press.
12 Sarafyazd, M.; Jazayeri, M. (2019) Hierarchical Reasoning by Neural Circuits in the Frontal Cortex. *Science* 364, 652. doi: 10.1126/science.aav8911.
13 Berger, Kevin. (2019) Talking Is Throwing Fictional Worlds At Each Other. *Nautilus,* 76. Retrieved from https://nautil.us/issue/76/language/-talking-is-throwing-fictional-worlds-at-one-another.
14 Koestler, A. (1964) *The Act of Creation.* 430. London: Hutchinson & Co.

15 Salthe, S. N. (1985) *Evolving Hierarchical Systems: Their Structure and Representation.* 61. New York: Columbia University Press.

Chapter 5

1 Lovejoy, Arthur O. (2009) *The Great Chain of Being: A Study of the History of an Idea.* p 22. New Brunswick : Transaction Publishers. A reprint of the original 1936 publication with a new Introduction by Peter J. Stanlis.
2 Lewes, G. H. (1879) *Problems of Life and Mind, Third Series* (Vol. Problem the First: The Study of Psychology). 179. London: Trubner & Co., Ludgate Hill.
3 Gefter, Amanda. (2015, November 19) Quantum Mechanics is Putting Human Identity on Trial. *Nautilus,* 30, . Retrieved from http://nautil.us/issue/30/identity/quantum-mechanics-is-putting-human-identity-on-trial.
4 Mach, Ernst. (1894) *Popular Scientific Lectures.* [trans.] Thomas J. McCormack. Third Ed. 205. Chicago : The Open Court Publishing Co.
5 Cole, K.C. (2019, July 17) The Simple Idea Behind Einstein's Greatest Discoveries. *Quanta.* Retreived from https://www.quantamagazine.org/einstein-symmetry-and-the-future-of-physics-20190626/
6 Englert, F.; Brout, R. (1964) Broken symmetry and the mass of gauge vector mesons. *Physical Review Letters,* 13 (9), 321–323. doi:10.1103/PhysRevLett.13.321. Higgs, Peter W. (1964) Broken symmetries and the masses of gauge bosons. *Physical Review Letters* 13 (16), 508–509. doi:10.1103/PhysRevLett.13.508. Guralnik, G.S.; Hagen, C.R.; Kibble, T.W.B. (1964) Global conservation laws and massless particles. *Physical Review Letters,* 13 (20), 585–587. doi:10.1103/PhysRevLett.13.585.
7 Mayr, Ernst. (1982) *The Growth of Biological Thought: Diversity, Evolution, and Inheritance.* 67. Cambridge, Massachusetts: The Belknap Press of Harvard University Press.
8 Simard, S.W., Perry, D.A., Jones, M.D., Myrold, D.D., Durall, D.M., Molina, R. (1997) Net transfer of carbon between ectomycorrhizal tree species in the field. *Nature* 388, 579-582. doi:10.1038/41557.
9 Simard, S. (2016) How trees talk to each other. Retrieved from https://www.ted.com/talks/suzanne_simard_how_trees_talk_to_each_other .
10 Strogatz, Steven. (2003) *SYNC: The Emerging Science of Spontaneous Order.* New York : Hyperion.
11 Thompson, J. Arthur. (1908) *The Bible of Nature: Five Lectures Delivered Before Lake Forest College on the Foundation of the Late William Bross.* (n.p.): (n.p.)..

Chapter 6

1 Anderson, P.W. (1972) More is Different: Broken Symmetry and the Nature of the Hierarchical Structure of Science. *Science* 177, 393-396. Retrieved from https://science.sciencemag.org/content/sci/177/4047/393.full.pdf.
2 Smith, Robin, "Aristotle's Logic", The Stanford Encyclopedia of Philosophy (Fall 2020 Edition), Edward N. Zalta (ed.), Retrieved from https://plato.stanford.edu/archives/fall2020/entries/aristotle-logic/.
3 "Physics (Aristotle)" (2019, Dec. 28) Wikipedia. Retrieved from https://en.wikipedia.org/wiki/Physics_(Aristotle)#The_meaning_of_physics_in_Aristotle .
4 Nagel, E. (1961) *The Structure of Science. Problems in the Logic of Explanation.* 352. New York : Harcourt, Brace & World, Inc.
5 Hopkins, E.J., Weisberg, D.S., Taylor, J.C.V. (2016) The seductive allure is a reductive allure: People prefer scientific explanations that contain logically irrelevant reductive information. *Cognition* 155, 67-76. doi: 10.1016/j.cognition.2016.06.011.
6 Needham, J. (1936) *Order and Life.* p 7. New Haven: Yale University Press.
7 Lewes, George Henry. (1879) *Problems of Life and Mind, Third Series, Vol. Problem the First: The Study of Psychology.* 179. London : Trubner & Co., Ludgate Hill.

Chapter 8

1 Bourne, M. (2018, March 6) *The Area Under a Curve.* Retrieved from https://www.intmath.com/integration/3-area-under-curve.php.
2 Kant, Immanuel. (1763) *Metaphysical Foundations of Natural Science.*
3 Friedman, Michael. (2013) *Kant's Construction of Nature.* Cambridge: Cambridge University Press, See the review at http://news.stanford.edu/2015/08/25/kant-newton-friedman-082515/

Chapter 9

1 Harris, S., Goldstein, R., Tegmark, M. (2019, Feb. 25) *What Is and What Matters.* Retrieved from https://www.youtube.com/watch?v=cypm7hkJ2lQ.

Chapter 10

1 Bornstein, A. M. (2016, Sept. 1) Is Artificial Intelligence Permanently Inscrutable? *Nautilus,* 40. Retrieved from http://nautil.us/issue/40/learning/is-artificial-intelligence-permanently-inscrutable.

Chapter 11

1 Cole, Kenneth S. (1968) *Membranes, Ions and Impulses: A Chapter of Classical Biophysics.* 241. Berkeley and Los Angeles: University of California Press.
2 Hodgkin, A.L., Huxley, A.F., Katz, B. (1952) Measurement of Current-Voltage Relations in the Membrane of the Giant Axon of *Loligo. J. Physiol.* 116, 424-448. doi:10.1113/jphysiol.1952.sp004716.
3 Hodgkin, A. L., Huxley, A. F. (1952) A Quantitative Description of Membrane Current and Its Application to Conduction and Excitation in Nerve. *J. Physiol.* 117, 500-544. doi:10.1113/jphysiol.1952.sp004764.
4 Fermi, E., Pasta, J., Ulam, S. (1955) *Studies of Nonlinear Problems I.* Los Alamos, New Mexico: Los Alamos National Laboratory.

Chapter 12

1 Johnson, E.A., Kootsey, J.M. (1985) A Minimum Mechanism for Na+ - Ca++ Exchange. *J. Memb. Biol.*, 86:167-187. doi:10.1007/bf01870783.
2 Brent, R. P. (1973) *Algorithms for Minimization Without Derivatives.* Englewood Cliffs, New Jersey: Prentice-Hall, Inc.
3 Strang, G. (1986) *Introduction to Applied Mathematics.* 665. Wellesley, Massachusetts: Wellesley-Cambridge Press.
4 Gleick, James. (1987) *Chaos: Making a New Science.* 9. New York: Penguin Books.
5 Murray, J.D. (1989) *Mathematical Biology.* 5. Berlin: Springer-Verlag.
6 Ibid. 45.
7 Ibid. 45.
8 May, Robert M. (1976) Simple Mathematical Models with Very Complicated Dynamics. *Nature* 261:459-467.
9 Ricklefs, Robert E., Miller, Gary L. (2000) *Ecology (Fourth Edition).* 353. New York: W. H. Freeman and Company.

Chapter 13

1 Pagels, Heinz R. (1988) *The Dreams of Reason: The Computer and the Rise of the Sciences of Complexity.* New York : Simon and Schuster.
2 Ibid. 45.
3 Ibid. 43.
4 Ibid. 53.
5 R.P. (1992) The Third Branch of Science Debuts. *Science* 256:44-47. doi:10.1126/science.256.5053.44

6 Cerf, V. G. (2016) Heidelberg Anew. *Comm. ACM* 59, 11, 7. doi.org/10.1145/3005354.
7 Horgan, J. (1997) *The End of Science: Facing the Limits of Knowledge in the Twilight of the Scientific Age.* New York: Broadway Books.
8 Penrose, Sir Roger. (1989) *The Emperor's New Mind: Concerning Computers, Minds, and the Laws of Physics.* Oxford: Oxford University Press.
9 Stent, G.S. (1969) *The Coming of the Golden Age: A View of the End of Progress.* Published for the American Museum of Natural History [by] the Natural History Press.

Chapter 14

1 Lewes, George Henry (1874) *Problems of Life and Mind, First Series, Volume 1: The Foundations of a Creed.* 174. Boston: James R. Osgood and Company.
2 Clayton, Philip. (2004) *Mind & Emergence: From Quantum to Consciousness.* 9. Oxford : Oxford University Press.
3 Carroll, Sean. (2016) *The Big Picture,* eChapter 12. New York: Dutton, an Imprint of Penguin Random House LLC.
4 Wilson, Edward.O. (1998) *Consilience: The Unity of Knowledge.* 94. New York : Vintage Books, a division of Random House, Inc.
5 Holland, John H. (1998) *Emergence: From Chaos to Order.* New York: Basic Books.
6 Morowitz, Harold .J. (2002) *The Emergence of Everything: How the World Became Complex.* Oxford : Oxford University Press.
7 Ridley, Matt. (2015) *The Evolution of Everything: How New Ideas Emerge.* New York : HarperCollins Publishers, Inc.
8 Roshanzamir, Ali (2013). "Matematik-professoren leger med lego-klodser". University of Copenhagen Faculty of Science. Archived from the original on 2 April 2015.
9 Morris, Simon Conway, ed. (2008) *The Deep Structure of Biology: Is Convergence Sufficiently Ubiquitous to Give a Directional Signal?* West Conshohocken, Pennsylvania: Templeton Foundation Press.

Chapter 15

1 Andreasen, Nancy C. (2014, July/August) Secrets of the Creative Brain. *The Atlantic,* 62.
2 Fernyhough, Charles. (2017, August) Talking to Ourselves. *Sci. Am.,* 74.
3 Gladwell, Malcolm. (2011) *Outliers: The Story of Success.* New York: Little, Brown and Company. 41.

4 Nightingale, Rob. (2011) The 10,000 Hour Rule is Wrong. How to Really Master a Skill. Retrieved from https://www.makeuseof.com/tag/10000-hour-rule-wrong-really-master-skill/ .
5 Koestler, Arthur. (1964) *The Act of Creation.* Hutchinson & Co. (Penguin Random House).

Chapter 16

1 Berryman, Alan A. (1992) The Origins and Evolution of Predator-Prey Theory. *Ecology* 73(5), 1530-1535. doi:10.2307/1940005.
2 Strogatz, S. (2003) *Sync: The Emerging Science of Spontaneous Order.* New York: Hyperion Books.
3 Simard, S.W., Perry, D.A., Jones, M.D., Myrold, D.D., Durall, D.M., Molina, R. (1997) Net transfer of carbon between ectomycorrhizal tree species in the field. *Nature* 388, 579-582. https://doi.org/10.1038/41557.
4 Simard, S. (2016) How trees talk to each other. Retrieved from https://www.ted.com/talks/suzanne_simard_how_trees_talk_to_each_other.
5 Greenwood, Veronique. (2017, November 30) How Bacteria Help Regulate Blood Pressure. *Quanta Magazine.* Retrieved from https://www.quantamagazine.org/how-bacteria-help-regulate-blood-pressure-20171130/ .
6 Bodin, Ö. Alexander, S.M., *et al.* (2019) Improving network approaches to the study of complex social-ecological interdependencies. *Nature Sustainability* 2:551-559.
7 Pennisi, E. (2015) Africa's soil engineers: Termites. *Science* 347, 596-597. doi:10.1126/science.347.6222.596.
8 Bonachela, J.A., Pringle, R.M., Sheffer, E., Coverdale, T.C., Guyton, J.A., Caylor, K. K., Levin, S.A., Tarnita, C.E. (2015) Termite mounds can increase the robustness of dryland ecosystems to climatic change. *Science* 347, 651-655. doi:10.1126/science.1261487.
9 Fredrickson, James K. (2015) Ecological communities by design. *Science* 348, 1425-1427. doi:10.1126/science.aab0946.
10 Teague, Brian P., Weiss, Ron. (2015) Synthetic communities, the sum of parts: Complex behaviors are engineered from cooperating cell communities. *Science* 349, 924-925. doi:10.1126/science.aad0876.
11 Sales-Pardo, Marta. (2017) The importance of being modular. *Science* 357, 128-129. doi:10.1126/science.aan8075.
12 Gilarranz, L. J., Rayfield, B., Linan-Cembrano, G., Bascompte, J., Gonzalez, A. (2017) Effects of network modularity on the spread of perturbation impact in experimental metapopulations. *Science* 357, 199-201. doi:10.1126/science.aal4122.

13 Yong, Ed. (2013, April) As One: How the Astonishing Power of Swarms Can Help Us Fight Cancer, Understand the Brain, and Predict the Future. *Wired*, 104-111, 136.
14 Winter, Dylan. (2010, November 13) Amazing Starlings Murmuration. Retrieved from https://www.youtube.com/watch?v=eakKfY5aHmY.
15 Atherton, Kelsey D. (2017) The Pentagon's new drone swarm heralds a future of autonomous war machines. Retrieved from https://www.popsci.com/pentagon-drone-swarm-autonomous-war-machines.
16 Easley, D., Kleinberg, J. (2010) *Networks, Crowds, and Markets: Reasoning about a Highly Connected World.* Cambridge University Press.
17 Klarreich, Erica. (2013, January 29) Computer Scientists take Road Less Traveled. *Quanta Magazine.* Retrieved from https://dev.quantamagazine.org/computer-scientists-find-new-shortcuts-to-traveling-salesman-problem-20130129/
18 Kim, Mark H. (2017, October 5) One-Way Salesman Finds Fast Path Home. *Quanta Magazine.* Retrieved from https://www.quantamagazine.org/one-way-salesman-finds-fast-path-home-20171005/.

Chapter 17

1 Harari, Y.N. (2015) *Sapiens: A Brief History of Humankind.* New York: HarperCollins Publishers.
2 Ibid., Chapters 10 and 16.
3 Bhandari, D.R. (1998, August) Plato's Concept of Justice: An Analysis. Retrieved from https://www.bu.edu/wcp/Papers/Anci/AnciBhan.htm.
4 Ariely, Dan; Garcia-Rada, Ximena. (2019, September) Contagious Dishonesty. *Scientific American*, 62-66.
5 Beirich, H, Buchannan, S. (2018, Spring) The Year in Hate and Extremism. *Intelligence Report (Southern Poverty Law Center),* 164, 33-62.
6 Wilson, E. O. (1998) *Consilience: The Unity of Knowledge.* 274. New York: Vintage Books, Random House, Inc.
7 Harris, Sam. (2010) Video: Science Can Answer Moral Questions. Retrieved from https://www.ted.com/talks/sam_harris_science_can_show_what_s_right
8 Goldstein, Rebecca N. (2017, December 11) The Scientists and the Philosophers Should Be Friends. *Free Inquiry,* 38 (1).
9 Vogel, G. (2004) The Evolution of the Golden Rule. *Science* 303, 1128-1131. doi:10.1126/science.303.5661.1128.
10 Hamilton, W. D. (1964) The Genetical Evolution of Social Behavior. *J. Theor. Biol.* 7, 1-16. https://doi.org/10.1016/0022-5193(64)90038-4.
11 de Vladar, H. P., Szathmary, E. (2017) Beyond Hamilton's rule: A broader view of how relatedness affects the evolution of altruism is emerging. *Science* 356, 485-486. doi:10.1126/science.aam6322.

12 Trivers, R. (1971). The Evolution of Reciprocal Altruism. *The Quarterly Review of Biology*, 46(1), 35-57. Retrieved from www.jstor.org/stable/2822435.
13 Darwin, C. (1871) *The Descent of Man*. London: John Murray.
14 Mayr, Ernst. (1988) *Toward a New Philosophy of Biology: Observations of an Evolutionist*. Cambridge, Massachusetts: The Belknap Press of Harvard University Press.
15 Bregman, Rutger. (2020) *Humankind: A Hopeful History (English version)*. 4. New York: Little Brown and Company.
16 Ibid., 21.
17 Marsh, Abigail A., *et al.* (2014) Neural and cognitive characteristics of extraordinary altruists. *PNAS* 111(42), 15036-15041. https://doi.org/10.1073/pnas.1408440111.
18 Chang, Steve W.C., *et al.* (2015) Neural mechanisms of social decision-making in the primate amygdala. *PNAS* 112(52), 16012-16017. https://doi.org/10.1073/pnas.1514761112.
19 Bhattacharjee, Yudhijit. (2018, January) The Science of Good and Evil. *National Geographic*. 116-143.
20 Wikipedia. (2019, December 16) Prisoner's dilemma. Retrieved from https://en.wikipedia.org/wiki/Prisoner%27s_dilemma
21 Nowak, Martin A., and Highfield, Roger. (2011) *SuperCooperators: Altruism, Evolution, and Why We Need Each Other to Succeed.* New York: Free Press, A division of Simon & Schuster, Inc.
22 Svoboda, E. (2017, July 2) Can Microbes Encourage Altruism? *Quanta Magazine.* Retrieved from https://www.quantamagazine.org/can-microbes-encourage-altruism-20170629/
23 Hanifan, L. J. (1916) The rural school community center. *Annals of the American Academy of Political and Social Science* 67, 130-138. https://www.jstor.org/stable/1013498.
24 Putnam, R. D. (1995, January) Bowling Alone: America's Declining Social Capital. *Journal of Democracy*, 65-78.
25 Claridge, Tristan. (2004) Definitions of Social Capital. Retrieved from https://www.socialcapitalresearch.com/literature/definition.htm.
26 Mineo, Liz. (2017, April) Good genes are nice, but joy is better. *Harvard Gazette.* Retrieved from https://news.harvard.edu/gazette/story/2017/04/over-nearly-80-years-harvard-study-has-been-showing-how-to-live-a-healthy-and-happy-life/
27 Quoted in the paper in note 3.

Chapter 18

1 Wilbur, David W. (2010) *Power and Illusion: Religion and Human Need.* (Self-published).

2 Kirsch, Adam. (2019, April 29). Modernity, Faith, and Martin Buber. *The New Yorker.* Retrieved from https://www.newyorker.com/magazine/2019/05/06/modernity-faith-and-martin-buber.

3 Buber, Martin. (2019 edition). *I and Thou.* San Francisco: Blurb, Inc.,

4 Gerson, Michael. (2018, April). The Last Temptation (How Evangelicals Lost Their Way). *The Atlantic*, 42-52.

5 Numbers, Ronald L. (2006) *The Creationists: From Scientific Creationism to Intelligent Design* (Expanded Edition). Cambridge, Massachusetts: Harvard University Press.

6 Johnson, Phillip E. (1995) *Reason in the Balance: The Case Against Naturalism in Science, Law and Education.* Downer's Grove, Illinois: InterVarsity Press.

7 Dembski, William A. (1999) *Intelligent Design: The Bridge Between Science and Theology.* Downer's Grove, Illinois: InterVarsity Press.

8 Dembski, William A. (2002) *No Free Lunch: Why Specified Complexity Cannot be Purchased without Intelligence.* Lanham, Maryland: Rowman & Littlefield Publishers, Inc.

9 Woodward, Thomas. (2003) *Doubts about Darwin: A History of Intelligent Design.* Grand Rapids, Michigan: Baker Books.

10 Meyer, Stephen C. (2009) *Signature in the Cell: DNA and the Evidence for Intelligent Design.* New York: HarperOne, an imprint of HarperCollins Publishers.

11 Johnson, Dominic. (2016) *God is Watching You: How the Fear of God Makes Us Human.* Oxford University Press.

12 http://sssreligion.org/about/history/

13 Fitzgerald, Michael. (2017, February 17) Atheism, the Computer Model: Big data meets history to forecast the rise and fall of religion. *Nautilus*, 45. Retrieved from http://nautil.us/issue/45/power/atheism-the-computer-model .

14 Wildman, Wesley. (2018, June 11) Religion is uniquely human, but computer simulations may help us understand religious behavior. *The Conversation.* Retrieved from https://theconversation.com/religion-is-uniquely-human-but-computer-simulations-may-help-us-understand-religious-behavior-79826 .

15 Asma, Stephen T. (2018) *Why We Need Religion.* New York: Oxford University Press.

16 Volf, Miroslav. (2015) *Flourishing: Why We Need Religion in a Globalized World.* New Haven, Connecticut: Yale University Press.

17 Numbers, Ronald L., ed. (2010) *Galileo Goes to Jail and Other Myths about Science and Religion.* Harvard University Press.

18 Video: "Bill Nye Debates Ken Ham." Retrieved from https://www.youtube.com/watch?v=z6kgvhG3AkI.

19 Gould, Stephen Jay. (2003) *The Hedgehog, the Fox, and the Magister's Pox: Mending the Gap Between Science and the Humanities.* 87. New York: Harmony Books.

20 Kauffman, Stuart A. (2008) *Reinventing the Sacred: A New View of Science, Reason and Religion.* Basic Books, New York.

21 Ibid., 284-285

22 Frank, Adam. (2009) *The Constant Fire: Beyond the Science vs. Religion Debate.* Berkeley and Los Angeles, California: University of California Press.

23 Ibid., 79.

24 Ibid., 254.

Chapter 19

1 Tucson, Reed V. (2020) The Disease of Distrust. *Science* 370 (6518):745. DOI: 10.1126/science.abf6109

2 "Measles Vaccination: Myths and Facts." (2019) Retrieved from https://www.idsociety.org/globalassets/idsa/public-health/measles/updated_-idsa_measlesvaccine_factsheet_4.25.19.pdf .

3 Wakefield AJ, Murch SH, Anthony A, Linnell J, Casson DM, Malik M, et al. (1998) Ileal-lymphoid-nodular hyperplasia, non-specific colitis, and pervasive developmental disorder in children. *Lancet* 351:637–41.

4 Pinker, Steven. (2018, Feb. 13) The Intellectual War On Science. *The Chronicle of Higher Education.* Retrieved from https://www.chronicle.com/article/The-Intellectual-War-on/242538.

5 Menczer, F., and Hills T. (2020, December) The Attention Economy: Understanding how algorithms and manipulators exploit our cognitive vulnerabilities empowers us to fight back, p 54. *Scientific American.*

6 Finkel, E. J., Bail, C.A., Cikara, M., Ditto, P.H., Iyengar, S., Klar, S., Mason, L., McGrath, M. C., Nyhan, B., Rand, D. G., Skitka, L.J., Tucker, J.L., Van Bavel, J. J., Wang, C. S., Druckman, J. M. *Science* 370 (6516), 533-536. DOI: 10.1126/science.abe1715.

7 Tsybeskov, Leonid. (2020, October) Science's Endangered Reputation, p 10. *Physics Today.*

8 *Brown Alumni Magazine*, Nov.-Dec., 2020, p12.

9 Oreskes, N., and Conway, E. M. (2011) *Merchants of Doubt: How a Handful of Scientists Obscured the Truth on Issues from Tobacco Smoke to Global Warming.* New York: Bloomsbury Press.

10 Oreskes, N. (2020, December) History Matters in Science, p 81. *Scientific American.*

11 Ritchie, Stuart. (2020) *Science Fictions: How Fraud, Bias, Negligence, and Hype Undermine the Search for Truth.* New York: Henry Holt and Company.

12 Reed, Genna. (2020, October) The Disinformation Pandemic, pp 11-12, *The Scientist.*

Chapter 20

1 In Letter to the Abbé Reynal, on the 'Affairs of North America in Which the Mistakes in the Abbé's Account of the Revolution of America are Connected and Cleared up', collected in the *works of Thomas Paine* (1797), Vol. 1, 295. Originally published in the Pennsylvania magazine (1775). Retrieved from https://todayinsci.com/P/Paine_Thomas/PaineThomas-Quotations.htm#:~:text=Every%20science%20has%20for%20its,he%20can%20only%20discover%20them.
2 Wilson, E. O. (1999) *Consilience: The Unity of Knowledge.* 13. New York: Vantage Books, A Division of Random House, Inc.
3 Feyerabend, Paul. (1995) *Against Method: Outline of an Anarchistic Theory of Knowledge.* London: New Left Books.
4 Goldstein, Rebecca N. (2017, December 11) The Scientists and the Philosophers Should Be Friends. Free Inquiry, 38 (1).
5 Tesla, Nikola. https://www.goodreads.com/author/quotes/278.Nikola_Tesla .

INDEX

D

E

F

G

H

I

J

K

L

M

N

O

P

Q

R

S

T

U

V

W

Printed in the United States
by Baker & Taylor Publisher Services